BIOLOGY
Science for Life
LABORATORY MANUAL

Virginia Borden   Colleen Belk

University of Minnesota, Duluth

PEARSON

Prentice
Hall

Upper Saddle River, NJ 07458

Assistant Editor: Colleen Lee
Executive Editor: Teresa Ryu Chung
Editor-in-Chief, Science: John Challice
Vice President of Production & Manufacturing: David W. Riccardi
Executive Managing Editor: Kathleen Schiaparelli
Assistant Managing Editor: Becca Richter
Production Editor: Dana Dunn
Supplement Cover Manager: Paul Gourhan
Manufacturing Buyer: Ilene Kahn
Photo Credits: Hang Glider - AP/Wide World Photos, Tomatoes - D. Cavagnaro/DRK Photos, Bee on Flower - David McGlynn/Getty Images, Inc., Brain - Volker Steger/Photo Researchers, Inc., Cell Division - P. Motta/Univ. "LaSapienza" Rome/Photo Researchers, Inc., Skeleton - Alfred Pasieka/Photo Researchers, Inc.

© 2005 Pearson Education, Inc.
Pearson Prentice Hall
Pearson Education, Inc.
Upper Saddle River, NJ 07458

Printed in the United States of America

10 9 8 7 6 5 4 3 2 1

ISBN 0-13-146917-7

Pearson Education Ltd., *London*
Pearson Education Australia Pty. Ltd., *Sydney*
Pearson Education Singapore, Pte. Ltd.
Pearson Education North Asia Ltd., *Hong Kong*
Pearson Education Canada, Inc., *Toronto*
Pearson Educación de Mexico, S.A. de C.V.
Pearson Education—Japan, *Tokyo*
Pearson Education Malaysia, Pte. Ltd.

Table of Contents

Preface

The *Biology: Science for Life Laboratory Manual* is designed to complement the textbook *Biology: Science for Life*. As with the textbook, the lab manual will actively engage you and develop your scientific literacy and critical thinking skills. The fifteen topics in the manual mirror the fifteen chapters of the textbook. In addition, because the same authors wrote both texts, the integration between the lecture and laboratory will be nearly seamless.

The elements that distinguish the textbook are repeated and extended in the laboratory manual. As in the text, the lab manual:

Demonstrates science as a process.   Each topic contains one or more exercises on hypothesis testing and evaluation of results. Throughout your use of the manual, your critical thinking skills will be developed as you learn to appreciate the explanatory strength of the scientific method. In addition, because many of the hypothesis tests in the manual are open-ended and do not produce "canned" results, you will understand the challenges of science and begin to see the dynamic nature of biologists' search for an understanding of the natural world.

Uses learning objectives to outline key points.   Each laboratory topic in the manual contains a list of learning objectives that describe the learning goals for the exercises within the topic. As with the other *Biology: Science for Life* resources, these learning objectives are closely tied to the key concepts found in the associated textbook chapters. All of the exercises in each lab are designed to meet one or more of these learning objectives. Additionally, because the labs explicitly state objectives, you will be better prepared to focus on the essential points of the lab.

Engages through analogies and models.   Most labs contain exercises that describe complex biological processes using simple and familiar items. These analogies will allow you to focus on the essential processes you need to understand without getting bogged down in the more technical details of the process. Many of these exercises are followed up by a more traditional lab activity. By learning via analogy first, you will be better able to appreciate the meaning and usefulness of the more traditional activities.

Relates science to everyday topics.   As in the textbook, each lab topic is connected to additional topics that have both scientific and social aspects. Incorporating larger societal questions into the laboratory setting will allow you to appreciate the usefulness of science to everyday life, and will also serve to humanize both the process of science and its practitioners. In large classes, laboratory sections can be the best places for substantive discussion – and consequently, substantive learning.

Writing is never a solitary endeavor. We are thankful for our excellent editors, Colleen Lee and Teresa Chung, whose support during the production of the manual was invaluable. Sincere thanks also go out to Dana Dunn and Becca Richter who skillfully coordinated the entire production process. Most importantly, we owe a debt of thanks to our reviewers, listed below. This lab manual has been much improved by their critical review and excellent suggestions. We are extremely grateful for their efforts.

Sylvester Allred   Northern Arizona University
Lynne Arneson   American University
Donna Becker   Northern Michigan University
David Byres   Florida Community College - Jacksonville
Danette Carlton   Virginia State University
William Coleman   University of Hartford
Heather DeHart   Western Kentucky University
Elizabeth A. Desy   Southwest Minnesota State University
William J. Edwards   Niagara University
Anne Galbraith   University of Wisconsin – La Crosse
Tony J. Greenfield   Southwest Minnesota State University
Andrew Goyke   Northland College
Julie Hens   University of Maryland, University College
Leland N. Holland   Pasco Hernando Community College
Hetty B. Jones   Savannah State University
Michelle Mabry   Davis and Elkins College
Ken Marr   Green River Community College
Monica McGee   University of North Carolina – Wilmington
Elizabeth McPartlan   De Anza College
John McWilliams Oklahoma Baptist University
Diane Melroy   University of North Carolina – Wilmington
Marjorie B. Miller   Greenville Technical College
James Munger   Boise State University
Greg Pryor   University of Florida
Pele Eve Rich   Mt. San Jacinto College
Susan Rohde   Triton College
John Richard Schrock   Emporia State University
Joanne Russell   Manchester Community College
Julie Schroer   Bismarck State College
Douglas Smith   Clarion University of Pennsylvania
Carol St. Angelo   Hofstra University
Janet Vigna   Grand Valley State University
James A. Wallis   St. Petersburg College
Lisa Weasel   Portland State University
Judy Williams   Southeastern Oklahoma State University
Danette Young   Virginia State University

Virginia Borden and Colleen Belk

TOPIC 1

The Scientific Method

Learning Objectives

1. Describe science as a process of proposing and testing hypotheses.
2. Distinguish between statements that are testable by science and those that are not.
3. Describe the meaning and purpose of experimental control.
4. Design an experiment to test a hypothesis.
5. Define "statistical significance" and describe what statistics describing an experimental result can and cannot tell us about the hypothesis.
6. Distinguish between "statistical significance" and "practical significance."

Pre-laboratory Reading

The word "science" is derived from a Latin verb *scientia*, meaning, "to know." Science is a way of knowing about the world. There are many other ways of knowing—in other words, of finding truth—including faith, philosophy, and cultural tradition. All of these ways of knowing help us understand different aspects of our world.

The essence of science as a way of knowing is the formulation and testing of statements called **hypotheses**. A hypothesis is a tentative explanation of how something works or of the cause of an event. Scientific hypotheses must be testable. In other words, hypotheses must be formulated in such a way that unambiguous observations of the natural world can help us support or reject them. Hypotheses are tested via **objective** observations; that is, observations that when made by many different, independent people would produce the same results. This helps us understand why not all explanations of "how the world works" are scientific hypotheses. Non-scientific hypotheses include opinions, hypotheses with statements that could be interpreted differently by different people, and hypotheses that invoke a **supernatural** cause or event. Supernatural entities or forces do not conform to the laws of nature and thus cannot be objectively observed; therefore, hypotheses that rely on the super-natural cannot be scientific.

Testing Hypotheses

Many cleansing products on the market today advertise that they "kill germs," or more specifically, that they are "antibacterial." This label typically indicates that these products contain triclosan or another related antiseptic chemical. Consumers concerned about food poisoning or other bacteria-caused illnesses preferentially purchase these products based on their assumption that the following hypothesis is true:

Antibacterial soap kills more bacteria on hands than standard hand soap.

Many scientific hypotheses are very logical. The hypothesis about antibacterial soap is one of these; after all, if manufacturers put antiseptic in soap it should kill more bacteria than soap without antiseptic, right? However, until the hypothesis is subjected to a test, it remains tentative. Good scientists are usually skeptical of untested hypotheses. For instance, there are reasons to doubt the hypothesis that antibacterial soap is a more effective cleanser. The action of hand washing *physically* removes bacteria from hands. It is not clear that the chemical composition of the soap (outside of the fats, oils, and alkali that all soaps contain) is an important factor in disinfecting hands. Without testing the hypothesis, there is no way to determine whether it is correct.

The first step in testing a hypothesis is making a **prediction** about the observations one would expect to make if the hypothesis was correct. You can think of a prediction as the "then" part of an "If … then…" statement. In other words, "If this hypothesis is true, then I expect to observe…." A prediction forms the basis for evaluating the truth of any statement. The prediction of the hypothesis that antibacterial soap kills more bacteria on hands than standard hand soap is that *hands washed with antibacterial soap will have fewer bacteria on them than hands washed with standard soap.*

Many scientific hypotheses can be tested through experimentation. An **experiment** is a contrived situation set up by a researcher solely for the purpose of testing a hypothesis. The hypothesis about antibacterial soap is testable by setting up an experimental situation that allows us to evaluate the prediction of the hypothesis. Note however, that *not all scientific hypotheses can be tested by experimentation.* Many hypotheses about historical events, because these are ideas about what happened in the past, cannot be experimentally tested because conditions have changed. For example, we cannot use experiments to test the hypothesis that humans evolved from primate ancestors, that an asteroid strike caused the extinction of the dinosaurs, or that agriculture developed in the Middle East and its practices migrated from there throughout Europe and Asia. However, these hypotheses are still testable, by objective observations of the natural world. For instance, if the hypothesis that agriculture developed first in the Middle East is correct, we would predict that the oldest agricultural implements discovered by archeologists would be found there.

When a hypothesis can be tested through experimentation, the most effective way to remove ambiguity from analysis of the results is to design a controlled experiment. **Control** indicates that the researcher works to ensure that all subjects in the experiment are treated identically (except for the experimental treatment) and that no consistent differences exist between individuals exposed to an experimental treatment and individuals who are not. In other words, a control helps to verify that the effect of an experimental treatment is due to the treatment itself and not another factor. One common experimental technique used to eliminate differences between groups in an experiment is to keep both the subjects and the technicians performing the experiment unaware of which individuals are receiving the experimental treatment and which are not. Experiments designed in this manner are called **double blind**, because the participants cannot "see" what outcome is expected.

Measurements collected from tests of hypotheses are called **data**. Data collected from experimental and control groups are summarized and compared using the tool of **statistics**. Statistics is a specialized branch of mathematics designed to help scientists relate the results of a limited experiment to a larger population. Experiments test the effect of a factor on a small subpopulation–the experimental group. This experimental group is a **sample** of the population it is derived from, as is the control group in the experiment. A sample is always an imperfect reflection of the whole population, and occasionally, samples contain individuals who are very different from the "average" individual in the population. As a result of this possibility, there are two reasons that experimental and control samples may differ: (1) the experimental treatment has a real effect; or (2) by chance, individuals selected for the experimental group are quite different than the individuals selected for the control group. If the mathematical calculations of a statistical test indicate that there is a very low chance of the latter event, the results of an experiment are termed **statistically significant**. Statistically significant results provide support for a hypothesis.

LAB EXERCISE 1.1

Practice Identifying and Creating Scientifically Testable Hypotheses

A. Review these statements with your laboratory partners and be prepared to share your answers with your laboratory instructor and/or other students.

B. For each statement, determine whether it is a scientific hypothesis as written, and if not, why not.

C. If the statement is not a scientific hypothesis, try to modify the statement so that it is testable.

D. Are any hypotheses impossible to test objectively?

1. It is wrong to perform medically unnecessary cosmetic surgery.

2. Biology lab is more fun than a barrel of monkeys.

3. God created Earth and all living creatures.

4. Plants that are spoken to regularly grow more rapidly than plants that are not spoken to.

5. Women are more intelligent than men.

LAB EXERCISE 1.2

Preparing to Design an Experiment

In the introduction to the lab, we put forth the following scientific hypothesis:

Antibacterial soap kills more bacteria on hands than standard hand soap.

A. Discuss the following questions with your lab partners and be prepared to share your answers with the lab instructor and other students.

1. What objective measures could we use to test the hypothesis about the cleansing power of antibacterial soap?

2. If the hypothesis is correct, what would you predict the outcome of the test to be?

3. To test this hypothesis, we could simply survey everyone in lab regarding his or her use of this soap. Presumably, at least some of the class almost always uses antibacterial hand soap and some almost never use this type of soap. After we found these two classes of people, we could simply compare the number of bacteria found on their hands. Why is this a poor test of the hypothesis?

4. We could test the hypothesis by designating half the class as "antibacterial soapers" and the other half as "regular soapers," having everyone wash their hands with the soap they have been assigned to, and then comparing the number of bacteria found on the hands of members of each group. However, this approach also is flawed. Why?

LAB EXERCISE 1.3

Design and Perform a Controlled Experiment

The last discussion exercise should have led you to consider some of the factors you will need to control when testing the hypothesis that antibacterial soap kills more bacteria on hands than standard hand soap. Now you should be prepared to design a well-controlled experiment to test this hypothesis.

First, you will need a short primer on how bacteria levels can be counted.

- Bacteria are single-celled organisms that are much too small to be seen with the naked eye, and many can only be seen under the highest magnification of a typical light microscope.
- Bacteria reproduce rapidly when in contact with a nutrient source.
- If an individual bacterial cell is transferred to a gel-like nutrient source, the cell will multiply into millions of descendants, producing a colony of cells that is visible to the naked eye.
- Thus, the number of bacteria initially on a given surface can be estimated by transferring those bacteria to a petri dish filled with nutrient agar gel, giving those cells 24–48 hours to multiply, and then counting the number of visible colonies on the plate.

A. Work with your lab partners to design a controlled experimental test of the hypothesis.

Materials available:
- Liquid hand soap: one containing triclosan and a similar formula soap minus triclosan
- Petri dishes filled with nutrient agar (two per student)
- Sterile cotton swabs for transferring bacteria from hands to petri dishes
- Permanent markers

B. Write an outline of your experiment below. Be prepared to share this design with the lab instructor and/or your classmates.

C. When you have finished, your lab instructor will lead a discussion that will generate a class protocol based on the common and best elements of each plan.

D. Follow the class protocol to perform an experimental test of the hypothesis. Results will be available in the next class period.

LAB EXERCISE 1.4

Collect Experimental Results

A. Count the number of bacterial colonies on your agar plate and record them here:
  - Before washing _____
  - After washing _____
B. What was your treatment group? _____
C. Input your data into the Excel spreadsheet provided by the instructor.
D. Fill in Table 1.1, summarizing all of the data collected by the class. Recall that the number of colonies on the plates is approximately equivalent to the number of bacteria initially transferred.

TABLE 1.1

Student	Bacteria Transferred Before Washing	Bacteria Transferred After Washing with Soap X, OR…	Bacteria Transferred After Washing with Soap Y

LAB EXERCISE 1.5

Summarize Experimental Results

A. Calculate the average number of colonies in each treatment group by summing all values in a column and dividing by the number of students in each treatment group.
- Before _____
- Group X_____
- Group Y _____

B. Ask your lab instructor which soap was antibacterial and which was standard.
- Soap X _____
- Soap Y _____

C. Fill in Table 1.2.

TABLE 1.2

	Before Treatment	Washed with Antibacterial Soap	Washed with Standard Soap
Average number of colonies per plate			
95% confidence value (available from the Excel spreadsheet)			
Range of values (the lowest and highest values in the data set)			

D. Create a box and whiskers graph to summarize the data in the preceding table by using the template in figure 1.1
- The "95% confidence interval" is a measure of the variability of the data. It is essentially the range of values that has a 95% chance of containing the "true" population mean (in other words the average number of bacteria found if we had performed this experiment with the entire human population as our sample). The mean *minus* the 95% confidence value is the low end of this range, and the mean *plus* the 95% confidence value is the high end of this range. The true population mean has a 95% likelihood of being within the box on the graph.

LAB EXERCISE 1.6

Evaluate Results

A number of statistical formulas exist that would allow us to test whether the difference between the number of bacterial colonies produced by the two different hand-washing treatments is statistically significant. However, to avoid delving into a complicated discussion of statistics, we can use the visual

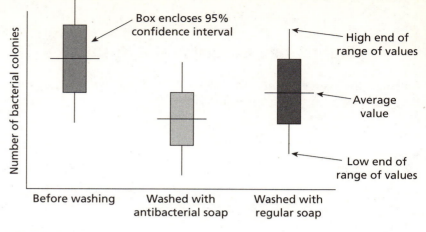

FIGURE 1.1

representation of the data in the graph that you just generated to determine whether the results are likely to show a significant difference.

In general, *if the confidence interval for either sample mean overlaps with the other sample mean, the average difference between the two sample means is not significant.* If this is the case, there is a more than 5% chance that the true population means of the two treatments are identical. The "5% chance" that there was no effect of the treatment is the standard value used by scientists to determine whether a difference between two samples is significant. In addition, the picture conveys other information:

- If the confidence interval boxes are large, the sample was extremely variable. This could indicate that a larger sample size is required to observe any effect of the experimental treatment.

- If the overlap between the confidence interval of one sample and the mean of another sample is relatively small, this suggests a difference between the two treatments. However, the results still do not allow us to reject the hypothesis.

A. Discuss the following questions with your lab partners and be prepared to share with your lab instructor and/or classmates.

   1. Examine the box and whiskers graph produced by the data. What does it tell you about the hypothesis that antibacterial soap kills more bacteria on hands than regular soap?

   2. Do you think that to get a clearer view of the status of the hypothesis, the experiment should be repeated with a larger sample size? Why or why not?

LAB EXERCISE 1.7

Discuss the Distinction Between Statistical Significance and Practical Significance

A statistically significant result does not necessarily tell us whether the results of the experiment indicate anything of practical value. For instance, a statistically significant but *small* difference between the number of bacteria on hands washed with antibacterial soap and hands washed with standard soap may have no effect on how healthy soap users are.

A. Discuss the following questions with your lab partners and be prepared to share your answers with the lab instructor and/or your classmates:

1. A drug that may soon become available to consumers has the following characteristics: It reduces the length of common cold symptoms (for most varieties of common cold) from 7.3 days to 6.3 days. Individuals who used the drug used 25% fewer tissues than people in the control group and had 33% fewer nights with disrupted sleep. These differences between the two groups are all statistically significant. What additional information about this drug, and what factors in your life, would influence whether these significant results were practically significant to you?

2. Sometimes, experimental results are statistically significant, but critics will argue that these results are an inconclusive test of the hypothesis. For instance, in a test of the hypothesis that regular aspirin consumption reduces the risk of heart attack, white male volunteers who took a single regular-strength aspirin each day had a 50% lower risk of heart attack than white males who did not take aspirin. Why would women and African American men argue that this hypothesis requires further testing?

TOPIC 1
_____
POST-LABORATORY QUIZ

THE SCIENTIFIC METHOD

1. Describe two characteristics that define a hypothesis as "scientific."

2. A hypothesis can be scientific and logical, but that does not mean it is true. To determine whether a hypothesis is likely to be true, it must be _____.

3. A recent experiment indicated that hormone replacement therapy in women increases the risk of suffering Alzheimer's disease. Hormone therapy is provided to women in pills containing estrogen and progesterone. What would be an appropriate control group for this experiment?

4. Scientists are interested in the following hypothesis: Exposure to bacteria and waste products present in and on farm and pet animals in early childhood reduces the risk of the child developing asthma or severe allergies. Describe a prediction of this hypothesis.

5. Define "double blind" in the context of an experiment.

6. Describe how you could experimentally test the hypothesis that ingesting the herb *Ginkgo biloba* improves memory. Be sure to include a description of the experimental control.

7. A relatively small group of individuals that is meant to be representative of a larger population is termed a _____ of that population.

8. A statistically significant difference between two samples is defined as one that has a _____ chance or less of occurring if the populations the two samples came from are not different.

9. What does a statistically significant result in an experiment comparing an experimental group to a control group mean?

10. The risk of premature death is 20% higher in smokers compared to nonsmokers. This difference is statistically significant and widely known. Using the idea of statistical versus practical significance, explain why so many people continue to smoke.

TOPIC 2
Nutrition and Metabolism

Learning Objectives

1. Describe the structure and function of enzymes.
2. Demonstrate that processing can denature enzymes.
3. Understand the role of mitochondria in producing cellular energy.
4. Demonstrate that carbon dioxide is produced during cellular respiration.
5. Understand how heart rate can be used as a measure of fitness.
6. Determine BMR and analyze a diet.

Pre-laboratory Reading

The food you eat goes through a series of conversions before it can be used to produce energy. When you swallow food, it moves from your mouth to your esophagus. The esophagus brings food to the stomach where the further breakdown of food into its subunits occurs. Food then travels through the small and large intestines. Substances that cannot be broken down exit the body after passing through the large intestine.

While in the stomach and small intestine, proteins called **enzymes** break the food down and release it to the blood stream where it is transported to individual cells. Enzymes function to speed up, or **catalyze**, the rate of metabolic reactions. Each enzyme catalyzes a particular reaction, a property called **specificity**. The specificity of an enzyme is the result of its shape. Different enzymes have different shapes because they are composed of a unique series of amino acids. The 20 amino acids each have different side groups, and are arranged in unique orders for each enzyme. This diversity in amino acid arrangement produces enzymes of various shapes and sizes. Enzymes can be broken apart down, or **denatured**, by heat or chemical treatments. After an enzyme has been denatured, it can no longer perform its cellular job.

Substances that are acted upon by enzymes are called the enzyme's **substrate**. The specificity of an enzyme for its substrate occurs because the enzyme can only bind to a substrate whose shape conforms to the enzyme's shape. The region on the enzyme where the substrate binds is called the enzyme's **active site**. The enzyme binds its substrate, helps convert it to a reaction product, and then resumes its original shape so it can perform the reaction again. A denatured enzyme loses its native shape and may no longer be able to bind its substrate.

After being broken down by the digestive system and delivered to individual cells, the products of digestion are used by organelles called **mitochondria**, located inside your body's cells (see Figure 2.1).

Mitochondria are kidney bean-shaped organelles, surrounded by two membranes, the inner and outer mitochondrial membranes. The inner membrane houses some of the proteins involved in producing ATP (adenosine triphosphate). The space between the two membranes is called the **intermembrane space**. Inside the inner membrane is the semifluid matrix of the mitochondrion where some of the enzymes involved in producing ATP are located. Once inside mitochondria, the nutrients in food undergo a process called cellular respiration.

During **cellular respiration**, cells use oxygen and produce carbon dioxide and water (this is why you breathe in oxygen and breathe out carbon dioxide). During this process, the energy stored in the chemical bonds of nutrients is used to produce ATP. The ATP produced can then be used to power cellular

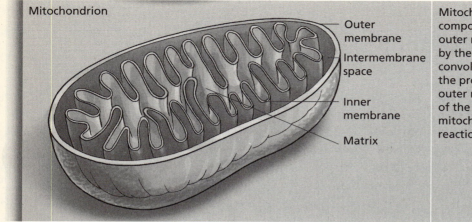

Mitochondrion

Outer membrane

Intermembrane space

Inner membrane

Matrix

Mitochondria are energy-producing organelles composed of two membranes. The inner and outer mitochondrial membranes are separated by the intermembrane space. The highly convoluted inner membrane carries many of the proteins involved in producing ATP. The outer membrane regulates traffic into and out of the mitochondrion. The matrix of the mitochondrion is the location of many of the reactions of cellular respiration.

FIGURE 2.1

activities, such as helping an enzyme perform its job. Cellular respiration occurs in the mitochondria of both plant and animal cells. The following equation summarizes the process of cellular respiration of a glucose molecule:

$$C_6H_{12}O_6 + 6O_2 \longrightarrow 6CO_2 + 6H_2O \ (+ATP)$$

glucose + oxygen yields carbon dioxide + water (+ energy)

Glucose is an energy-rich compound, but the products of its digestion—carbon dioxide and water-are energy poor. The energy released during the conversion of glucose to carbon dioxide and water is used to synthesize ATP.

Cellular respiration occurs at different rates in different individuals. When more calories are consumed than used, energy is stored as fat. When fewer calories are consumed than used, weight loss occurs. Your heart rate, in part, determines the rate of cellular respiration because the heart pumps oxygen to cells so that respiration can occur. The more efficient your heart is, the less work it has to do to supply your cells with oxygen.

LAB EXERCISE 2.1

Enzymatic Breakdown of Collagen

Canning is a type of processing that involves heating foods to high temperatures to kill any bacteria that might be present. In addition to killing microbes, proteins present in canned food can be denatured.

Collagen is a fibrous protein that serves as a structural subunit of bone, cartilage, and tendon. Boiling collagen converts the fibrous protein into a more gelatinous form. Collagen also serves as a substrate for the enzyme bromelin. Bromelin, present in plants called bromeliads, metabolizes collagen.

A. Witness the actions of the enzyme bromelin on collagen in gelatin.

  1. Devise a hypothesis that predicts how the condition of gelatin may be affected by exposure to fresh and canned pineapple juice.

2. Fill a beaker to 50 ml with fresh pineapple, cut in 1-inch chunks. After measuring, place the pineapple in a blender and puree for 1–2 minutes. Strain the puree through cheesecloth to remove the pulp.

3. Clean the blender and repeat this procedure using chunks of canned pineapple.

4. Add one 1/2-inch cube of gelatin to each of three test tubes. Press the gelatin to the bottom of the tube. Label the test tubes A, B, and C.

5. Pour 5 ml of freshly prepared pineapple juice into test tube A and 5 ml of juice from the canned pineapple into test tube B. Add nothing to test tube C.

6. What is the purpose of tube C?

7. Use Table 2.1 to record your observations every 5 minutes for 30 minutes. (You can perform Exercises 2 and 3 while these reactions are running.)

TABLE 2.1

	0 Minutes	5 Minutes	10 Minutes	15 Minutes	20 Minutes	25 Minutes	30 Minutes
Gelatin + fresh pineapple							
Gelatin + canned pineapple							

8. Do your results support your hypothesis? Why or why not?

LAB EXERCISE 2.2

Examining Mitochondria

After food has been digested in the stomach and small intestine, it is transported to cells for further breakdown. Mitochondria are kidney bean-shaped organelles found inside cells that help continue the breakdown process.

A. Stain plant cells to visualize mitochondria.

1. On a clean glass slide, mix three drops of the stain Janus Green B with one drop of 7% sucrose.

2. Prepare a very thin piece of red onion epidermis by removing a fleshy portion of the bulb and snapping the modified leaf backward. Look carefully at the point at which the onion snapped. Remove the thinnest section of onion that you can recover from the break point. Place the section in the stain you placed on the slide and add a cover slip.

3. Place the slide under the microscope, on low power, and focus on the edge of the section. The stained mitochondria will appear as small blue spheres. Rotate the nose-piece of the microscope to view on higher power.

4. Draw the onion cell and mitochondria you saw under the microscope.

LAB EXERCISE 2.3

Carbon Dioxide Production During Cellular Respiration

The mitochondria you viewed in the previous exercise use oxygen to convert food energy into the form of energy your cells can use, ATP. Cells undergoing this process release carbon dioxide. When cellular respiration occurs in your body, oxygen enters your cells when you breathe it in through your lungs and it is transported to your cells by red blood cells. The carbon dioxide that is produced is carried through the blood to the lungs and released.

The release of carbon dioxide can be measured by using an indicator solution called bromothymol blue. As carbon dioxide accumulates, the bromothymol blue turns yellow and is therefore an indicator of carbon dioxide concentration.

A. Measure carbon dioxide production at rest and with exercise.

1. In this exercise, you will measure carbon dioxide production at rest and with exercise. Propose a hypothesis about how carbon dioxide levels might change at rest and with exercise.

2. Place three drops of 0.004% bromothymol blue into each of three test tubes numbered 1–3.

3. One group member should be seated at rest for three minutes. At the end of the rest period, make a note of the time or start your timer. Have the rested group member exhale by gently blowing, at normal breathing intervals, through a drinking straw into the bromothymol blue in test tube 1. Time how long it takes for the solution to change color.

4. The same group member should do as many jumping jacks as possible in three minutes and then blow through the straw into the tube with each exhalation. Time how long it takes for the solution to change color in test tube 2.

5. Use a rubber bulb attached to the end of a Pasteur pipette to blow atmospheric air into the bromothymol in test tube 3 until there is a color change. How long did it take for a color change to occur?

6. Compare the results seen in test tubes 1–3. Do the results of this test support your hypothesis? Why or why not?

7. Why was it important that the same group member performed steps 3 and 4?

LAB EXERCISE 2.4

Heart Rate as Measure of Healthfulness

Aerobic exercise, exercise that uses oxygen, includes activities such as running, swimming, or biking. Aerobic exercise helps improve the fitness of your heart and its associated blood vessels, thereby improving your cardiovascular fitness. Measuring your heart rate during exercise can help you determine how hard your heart has to work to supply oxygen to your tissues for cellular respiration.

Knowing how to monitor your heart rate allows you to determine whether you are working out too intensely, which is dangerous for your heart, or not hard enough, which is less productive. For a given individual, a range of heart rates provides the most cardiovascular benefit. To maximize the benefits of your aerobic workout, you need to stay within this range of heart rates at least 20 to 30 minutes.

A. Determine your heart rate.

To determine your range of target heart rates, you must first measure your **resting heart rate**. Your resting heart rate is the rate your heart is pumping when you have been sleeping or sitting quietly for a while.

1. To estimate your resting heart rate, count the number of pulses at the carotid artery by pressing gently on one side of your neck, under your chin. You can count for 60 seconds to obtain the number of beats per minute (bpm) or for 6 seconds and multiply that number by 10. If you have been moving around during lab, subtract 15 bpm from the number you counted to estimate your actual resting heart rate. What is your estimated resting heart rate?

The typical adult has a resting heart rate of 60–80 bpm whereas highly trained athletes may have readings of 40 bpm or lower. As you become fit, your resting heart rate should decrease.

Your **maximum heart rate** is the highest number of beats per minute your heart should reach while exercising. This can be estimated by subtracting your age from 220.

2. What is your estimated maximum heart rate?

3. To calculate your target zone, multiply your maximum heart rate by 60% (.6) for the lowest heart rate in your target range and 80% (.8) for the highest heart rate in your target zone. What is your target zone?

As an athlete increases his cardiovascular fitness, he will be able to train for longer periods of time and more intensely without increasing his heart rate above the maximum for their target zone. This means that the athlete's heart is becoming stronger and more efficient.

LAB EXERCISE 2.5

Food and Energy

Just as different individuals have different target heart rates, different individuals also have different energy use efficiencies. The **metabolic rate** of an individual is a measure of his or her energy use. The metabolic rate changes according to an individual's activity level. The **basal metabolic rate (BMR)** is a measure of the energy used by an awake, alert person. This rate differs among individuals based on height, weight, age, and gender. You can estimate your BMR and use that information to calculate your daily caloric needs using a formula called the Harris Benedict formula.

Using the Harris Benedict formula to determine caloric needs is more accurate than measurements based solely on body weight. This is because two individuals who weigh the same amount might differ in exercise levels, age, height, and gender. An important factor that this formula does not take into account is body fat percentage, because this is often not known and is difficult to determine. Keep in mind that leaner, more muscular bodies need more calories because muscle is more energy costly than fat. Therefore, this equation is especially accurate for people who are not very muscular or very fat.

A. Determine the number of calories required to maintain your weight.

1. Determine your BMR.

$$\text{BMR for women} = 655 + (9.6 \times \text{weight in kilos}) + (1.7 \times \text{height in cm}) - (4.7 \times \text{age in years})$$

$$\text{BMR for men} = 66 + (13.7 \times \text{weight in kilos}) + (5 \times \text{height in cm}) - (6.8 \times \text{age in years})$$

Notes:

1 inch = 2.54 cm

1 kilogram = 2.2 lbs

For example Jean weighs 130 pounds and is 5′ 4″ tall.

Her weight in kilograms is 130 lbs $\times$ 1kg/2.2lbs $=$ 59 kg.

Her height in cm is 64 inches $\times$ 2.54 cm/inch $=$ 162.56 cm.

Your BMR $=$ _____

2. Multiply your BMR by a factor that describes your activity level.

   Sedentary activity (little or no exercise) $=$ BMR $\times$ 1.2

   Light activity (exercise 1–3 days/week) $=$ BMR $\times$ 1.375

   Moderate activity (exercise 3–5 days/week) $=$ BMR $\times$ 1.55

   Heavy activity (hard exercise 5–7 days/week) $=$ BMR $\times$ 1.725

   BMR $\times$ activity factor $=$ _____

This is the total number of calories you need to maintain your current weight.

B. Determine whether a particular diet is balanced.

   The percentage of calories in a healthy diet that is obtained from protein should be around 0.8 grams per 2.2 lbs. The percentage of calories from fat should be at least 10% but not more than 30%. The remaining calories should be provided by carbohydrates.

1. Helen is a 150-pound, 20-year-old college junior. A typical day's food for Helen consists of around 2,000 calories, composed of 60 grams of protein, 70 grams of fat, and 283 grams of carbohydrate. Calculate the percentage of calories provided by each nutrient (protein and carbohydrate are 4 calories/gram, fat is 9 calories/gram).

2. The only other things Helen should be concerned about are getting enough vitamins, minerals, fibers, and water. What types of foods would be most beneficial in this regard?

3. Track your own diet for one week and calculate the percentage of your diet that is fat, carbohydrate, and protein.

TOPIC 2
_____
POST-LABORATORY QUIZ

NUTRITION AND METABOLISM

1. Describe how enzymes catalyze metabolic reactions.

2. Different enzymes are composed of different orders of _____.

3. What does the enzyme bromelin do and why did this not occur when the canned pineapple was used?

4. List the reactants and products of cellular respiration.

5. The products of digestion are used to produce ATP in organelles called _____.

6. The _____ represents the resting energy use of an awake, alert person.

7. List several factors that affect a person's BMR:

8. Do proteins, carbohydrates, or fats have the most calories per gram?

9. Why is a fit person's heart rate slower than that of an unfit person?

10. If a very muscular person used the Harris Benedict formula to calculate daily calorie needs would the number of calories they calculated actually be an overestimate or an underestimate? Why? What if an overweight person used the Harris Benedict formula? Would they overestimate or underestimate their daily caloric needs?

TOPIC 3
Biodiversity

Learning Objectives
1. Describe the distinction between domain and kingdom.
2. Describe the characteristics of the major categories of living organisms.
3. Describe how the classification system appears to reflect the relationships among modern organisms.
4. Practice using a dichotomous key.
5. Gain an appreciation for the diversity of life on Earth.

Pre-laboratory Reading

The term **biodiversity** encapsulates the great variety of life—from the diversity of alleles (forms of genes) within a species, to the diversity of species that inhabit the planet. The diversity of life forms that surround us is part of what makes biology a fascinating and continually developing science. In this set of lab exercises, we review a tiny sample of the many forms of life on Earth. The exercises that make up this small sampling are too numerous for a single lab period—your laboratory instructor has had to make some tough decisions about how to cover this small sampling of biodiversity within the time constraints of your class time. To put into perspective the impossibility of surveying life in a single lab period, consider the following: If we were to spend 5 seconds viewing every species on Earth that has been identified (about 1.8 million), it would take us over three months of solid observation to see them all. And, if estimates about the actual number of species are correct (more than 200 million), it would take more than 30 years!

To organize the study of diversity, biologists have developed a **classification system** that categorizes organisms into similar groups. The largest group in the currently accepted classification system is called the **domain**. All living organisms can be classified into one of three great domains of life that are distinguished by fundamental differences in cell structure. The next largest group is the **kingdom**. Only one domain currently has well-defined kingdoms, which group organisms based on similarities in cell structure and mode of nutrition. The most widely used classification of living organisms divides life into three domains and four kingdoms. Our survey of diversity uses this classification system as a way to give a broad overview of the variety of life:

Domains

Bacteria Single-celled organisms without a nucleus and with cell walls containing **peptidoglycan**, a complex of protein and sugars. Although some bacteria cause disease, many are benign or beneficial to humans.

Archaea Single-celled organisms without a nucleus and with cell walls that do not contain peptidoglycan. Many of the known species grow in extreme environments, such as very salty or very high-temperature water.

Eukarya (see following kingdom descriptions) Organisms made up of cells with a membrane-bound nucleus that contains the genetic material of the cell. This is the only domain containing truly **multicellular** organisms; that is, those containing cells that are integrated and specialized for specific functions.

Kingdoms of Eukarya:

Protista (amoeba, algae, diatoms) All organisms in this group have a free-living single cell for at least part of their life stage. Most are unicellular, such as amoebae and paramecia, but many of the algae have a multicellular, stationary phase (such as kelp).

Plantae (plants) Multicellular organisms that are **autotrophic** (make their own food).

Fungi (fungi, molds, yeast) Multicellular organisms that are **heterotrophic** (rely on other organisms for food), digest food outside of their bodies, and absorb the products of digestion.

Animalia (animals) Multicellular organisms that are heterotrophic and ingest food for digestion inside the body.

The four kingdoms and two noneukarya domains are further classified into smaller groups as follows, from most- to least-inclusive:

Phylum
  Class
    Order
      Family
        Genus
          Species

For instance, in the kingdom Animalia, all species that produce an external skeleton are placed in the Phylum Arthropoda. Within this phylum, all species that have three major body segments (a head, thorax or "chest", and abdomen) are placed in the Class Insecta. Within the Insecta, all species that produce two hardened wings that form a protective covering for the two membranous wings beneath them are placed in the same order, Coleoptera—the beetles.

Our ability to group organisms into categories that reflect greater and greater degrees of similarity is considered evidence that living organisms are related to each other. In other words, the reason all species in the Order Coleoptera have similar wings is because they all derived from an ancestral species that had this type of wings. In fact, Charles Darwin used the pattern of similarity that drove the development of this classification system as one piece of evidence for the **theory of evolution**—the theory that all organisms alive today represent the descendents of a single ancestral species that arose on Earth billions of years ago. The theory of evolution is explored more thoroughly in Lab 8 in this manual.

One of the most serious environmental issues humans face today is the recent rapid decline of biodiversity, caused by human-induced extinctions of species from every class of organisms. Lab 14 in this manual explores the causes and consequences of the loss of biodiversity more thoroughly.

LAB EXERCISE 3.1

The Relationship Among the Domains of Life

If all modern organisms derived from a common ancestor in the past and if more similar species share a more recent common ancestor, we should be able to picture the relationships among living organisms as a tree containing a series of branching events. In this exercise, you will generate and test a hypothesis about the relationship between the three domains of life.

A. Examine the generic tree in Figure 3.1. This tree contains all of the major elements required when illustrating a hypothesis of evolutionary relationship among organisms.

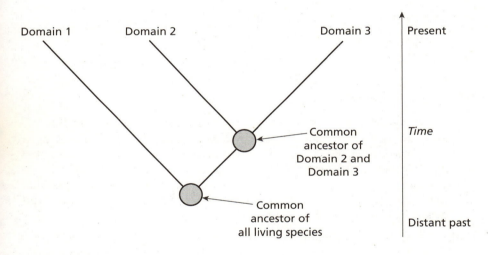

FIGURE 3.1

B. Discuss the following question with your lab partners and be prepared to share your answer:

1. Domain 2 and Domain 3 are hypothesized to share a more recent common ancestor with each other than either does with Domain 1. What observations would cause a scientist to make this hypothesis?

C. Given Table 3.1 of characteristics found in the three domains, work with your lab partners to create a hypothesis of evolutionary relationship among these groups. Basically, the more characteristics two domains share, the more likely they are to share a more recent common ancestor. Be prepared to share your hypothesis and to explain your reasoning.

TABLE 3.1

Characteristic	Bacteria	Archaea	Eukarya
Nucleus present	No	No	Yes
Cell contains internal structures surrounded by a membrane	No	No	Yes
Cell wall contains peptidoglycan	Yes	No	No
Type of RNA polymerase (that is, transcription enzyme)	One	Several	Several
Response to antibiotics streptomycin and chloramphenicol	Growth inhibited	Growth not inhibited	Growth not inhibited
Histone proteins associated with DNA	No	Yes	Yes
Circular chromosome	Yes	Yes	No
Genes contain DNA sequences that do not code for protein (that is, introns)	No	Some	Yes
Amino acid that initiates translation	Formyl-methionine	Methionine	Methionine
Ability to grow at temperatures exceeding 100° C	No	Some species	No

D. Test your hypothesis.

The hypothesis of evolutionary relationship represented by the tree you have just drawn can be tested via other data. In 1977, the biologist Carl Woese compared the DNA sequences of the gene that codes for part of the ribosome (the organelle that is responsible for translating DNA information into proteins) among a large number of species. The tree that Woese generated based on these DNA sequences is often called the "universal **phylogeny**" or universal family tree.

1. Use the resources available to you (for example, your textbook, the Internet, other biology texts) to locate a reproduction of Carl Woese's universal phylogeny.

2. Compare the tree you generated to Woese's tree. Does his tree support your hypothesis? Why or why not?

LAB EXERCISE 3.2

A Survey of the Domains Bacteria and Archaea

The diversity and importance of bacteria and archaea are often under appreciated because these organisms are so tiny. The first part of this exercise asks you to calculate the size of these cells and compare them to other objects.

A. Determine the size of a "typical" bacterial cell by following these steps:

1. Obtain a millimeter scale and place it on the microscope stage. Examine the scale with the scanning (3.5X or 4X) objective lens in place.

2. Estimate the diameter of the field of view by counting the number of 0.1 millimeter "blocks" that span the field of view. Record this diameter to the nearest 0.1 mm.

   Diameter of the scanning field _____

3. The diameter of the field of view using the scanning objective can be used to calculate the diameter using any other objective by the following formula:

$$\frac{\text{Magnification of scanning objective}}{\text{Magnification of "other" objective}} \times \text{diameter of scanning objective} =$$

Diameter of "other" objective _____

4. Calculate the diameter of the field of view of the highest power objective.

   Diameter of the high power field _____

5. Obtain a prepared slide of filamentous bacteria, and view it under the high power objective. Estimate the length and width of a single cell in the filament in millimeters using the following scheme:

Length of a strand that spans the field of view = diameter of field of view

$$\text{Length of a single cell} = \frac{\text{number of cells in strand spanning field of view}}{\text{length of the strand}}$$

Width of a single cell = number of "lengths" per width $\times$ single cell length

Length _____   Width _____

6. The diameter of a period on this page is approximately 0.1 mm. How many bacterial cells would be able to span a period? _____

B. Visualize the abundance of bacteria in the environment.

1. With a sterile cotton swab, obtain a sample of bacteria from your teeth. Gently rub the swab over the surfaces of several back teeth.
2. Carefully roll the swab across the surface of a clean glass slide. Try to spread the material thinly and evenly.
3. Place the slide across the top of the beaker and allow the smear to air dry.
4. Flood the smear with methanol for one minute or quickly pass the slide through a flame (right side up) two or three times. Allow the smear to air dry before staining.
5. Bring the slide and beaker to a sink and carefully flood the slide with crystal violet. This stain is messy—avoid contact with your clothes and skin and wipe up any spills thoroughly. Wait one to two minutes and then rinse the slide with cold tap water.
6. Gently shake off the excess water and allow the smear to air dry.
7. Observe the slide under the highest power microscope objective.
8. Estimate the number of bacterial cells that are visible in a field of view. _____
9. Estimate the size of the smear on the slide. _____
10. Calculate the number of bacteria you collected on the swab.

    Number of bacteria in field of view $\times$ size of smear = _____

C. Investigate the function of antibiotics.

Domain Bacteria contains many infamous members, including the organisms that cause tuberculosis, syphilis, gonorrhea, anthrax, cholera, tetanus, and leprosy, not to mention all manner of food poisoning and blood-borne infections. Yet in the larger picture, only a very small proportion of bacteria cause disease—most do not affect us at all or are beneficial to us.

The presence of **pathogenic** (disease-causing) bacteria and the severity of diseases they cause have driven interest in the development of drugs that selectively target and kill bacteria—generally called **antibiotics**. The presence of the unique compound peptidoglycan in bacterial cell walls provides a means for antibiotics to target bacterial cells without harming our own body cells. The antibiotic penicillin prevents molecules of peptidoglycan from forming a strong network and in many cases can greatly weaken a cell wall and thus kill the cell.

There is variation among bacterial species in their susceptibility to antibiotics. We can divide the bacterial domain into two major groups of bacteria: those with simple peptidoglycan cell walls and those with peptidoglycan cell walls surrounded by a membrane. The outer membrane on the latter type of bacteria impedes the movement of materials into the cells. As you might imagine, this type of bacteria tends to be less sensitive to antibiotics such as penicillin. However, broad-spectrum antibiotics exist that are toxic to both kinds of bacteria. These antibiotics tend to have more severe side effects in patients than the narrower-spectrum antibiotics.

1. Observe the two plates of bacteria available in lab. You should note the disks impregnated with antibiotic present on the surface of these plates. If the bacteria on the dish are susceptible to the antibiotic, a clear area will appear around the disk—if not, bacteria will grow up to the surface of the disk, causing a cloudy appearance. Fill in Table 3.2, noting which bacterial species are susceptible to which antibiotic:

TABLE 3.2

	Antibiotic A Name:	Antibiotic B Name:	Antibiotic C Name:	Antibiotic D Name:
Bacterial Species 1 Name:				
Bacterial Species 2 Name:				

2. Given what you have learned about the two groups of bacteria, make a hypothesis about which type of cell wall you expect to find in:

Bacterial Species 1 _____

Bacterial Species 2 _____

3. The two types of bacteria respond differently to a cell-staining procedure known as the Gram stain. You can test the hypothesis you made in step 2 by performing a Gram stain on the two bacterial species using the following protocol:

1. Obtain a clean glass slide and place a single drop of distilled water in the center of the slide.

2. Use a clean toothpick to obtain a tiny sample of the bacterial culture from one of the petri dishes.

3. Gently stir the toothpick in the water on the glass slide. Spread the water into a thin layer.

4. Place the slide on the beaker and allow it to air dry. Bring the slide and beaker to a sink for the staining process.

5. Flood the smear with methanol for 1 minute or pass the slide through a flame (right side up) two or three times. Allow the smear to dry before staining.

6. Flood the smear with crystal violet, and wait for 1–2 minutes. (Note: this and the other stains can permanently mar clothing. Avoid spills, and clean any promptly and thoroughly.)

7. Gently rinse off the crystal violet with cold tap water.

8. Flood the smear with iodine and wait for 1 minute.

9. Gently rinse off the iodine with cold tap water.

10. Rinse the smear with Gram Decolorizer until the solution rinses colorless from the slide (20–30 seconds).

11. Immediately rinse the smear with cold water.

12. Flood the smear with safranin and allow it to stain for 15–30 seconds.

13. Gently rinse with cold tap water.

14. Blot off excess water with a paper towel, and allow the smear to air dry.

15. Repeat the same procedure with the second bacterial culture.

16. Wash your hands thoroughly when finished.

    Bacterial cells that have the simple peptidoglycan cell wall will pick up the violet stain and will remain violet in color throughout the rest of the preparation—these cells are called Gram-positive bacteria. Those bacteria with the additional outer membrane outside the wall do not keep the violet stain throughout the procedure, but will stain with the second dye, the red safranin—these cells are called Gram-negative bacteria.

4. Examine the slides microscopically and record your observations in Table 3.3.

TABLE 3.3

	Gram positive or negative?
Bacteria Species 1	
Bacteria Species 2	

5. Do these observations support the hypothesis you made in step 2 of this exercise?

6. Which of the antibiotics used in this exercise appear to be broad-spectrum antibiotics?

LAB EXERCISE 3.3

A Survey of the Kingdom Protista

A. Observe various members of the Kingdom Protista.

   The protists are a mixed bag of highly diverse organisms. In general, we can group the members of this kingdom into three broad "functional groups": (1) the plant-like protists that make their own food, (2) the animal-like protists that are highly motile and engulf food, and (3) the fungus-like protists, which are often less motile and tend to dissolve their food source before absorbing its now-simplified nutrients.

   1. Examine the variety of protists available in the laboratory. Based on your observations, attempt to determine the basic life style of each and record your determinations in Table 3.4. Be prepared to share your answers (and your rationale) with your lab instructor and classmates.

TABLE 3.4

Protist	Plant-like, Animal-like, or Fungus-like?	What Is Your Evidence?

B. Investigate whether human activities in an environment affect the diversity of protists in that environment.

   1. Collect water and some sediment from two different ponds. One should be a storm water retention pond, such as those found near parking lots and other developments, while the other should be a natural pond in a park or protected area.

   2. Mix the water and sediment thoroughly and prepare a wet mount of each pond sample by placing a drop of liquid on a slide, adding a drop of methyl cellulose to slow down the movement of the organisms, and covering it with a coverslip.

   3. As best as you can, determine the number of different protistan species you can see in each drop of water. You do not need to identify these by their scientific name, simply keep track of how many different types you see. A sketch or brief description on the following table will help you in this task.

   4. Now count the number of individuals of each type you have identified and fill in Table 3.5.

TABLE 3.5

Pond 1:	Number Observed (A)	Number Observed − 1 (B)	A × B	Pond 2:	Number Observed (A)	Number Observed − 1 (B)	A × B
Species:				Species:			
Species:				Species:			
Species:				Species:			
Species:				Species:			
Species:				Species:			
Species:				Species:			
Species:				Species:			
Species:				Species:			
Species:				Species:			
Species:				Species:			
Species:				Species:			
Sum of Column (A × B)	███████			Sum of Column (A × B)	███████		

5. Complete the following calculations of the "diversity index" for each pond site and fill in Table 3.6:

$$\frac{\text{Total number of species seen} \times (\text{Total number of species seen} - 1)}{\text{Sum of Column (A} \times \text{B)}}$$

TABLE 3.6

	Diversity Index
Pond 1:	
Pond 2:	

6. The diversity indices in the preceding table allow you to compare the diversity in the two ponds as measured both by number of species (a measure called **species richness**) and abundance of species. By dividing a measure of the number of species you observed in each environment by a measure of the abundance of each species observed, you will generate a number that tells you whether one or a few species dominates in

that environment. Basically, having a large number of species leads to a greater diversity, but if nearly every individual in an environment belongs to a single species and there are only a few individuals in the remaining species, the environment is actually not very diverse. A larger diversity index equates to greater diversity.

7. Describe your results. Which environment has a more diverse protistan fauna? Why do you think this is the case?

LAB EXERCISE 3.4

A Survey of the Kingdom Animalia

A. Discuss the diversity of the animal kingdom.

Most people's conception of what constitutes an "animal" is surprisingly limited. Recent classifications identify 35 animal phyla, but nearly all animals we think of belong to three or four of these.

1. Keeping in mind the basic definition of "animal"—multicellular organisms that make their living by ingesting other organisms and which are motile during at least one stage of their life cycle—brainstorm a list of animals with your lab partners. Be prepared to share this list with your lab instructor and classmates.

2. With the help of your lab instructor, determine the number of different animal phyla represented on your list. How many phyla are represented by the class's list?

B. Learn the characteristics of some common animal phyla and practice using a dichotomous key.

Animals are grouped into phyla based on shared body form, skeleton type, and digestive system. We assume that animals present today that are similar in these three traits are all descendants of a single species that had these same traits.

1. Examine the models and diagrams available in the lab to understand the differences in body form, skeleton type, and digestive system (described next) that are found in modern animals.

Body forms are typically described in terms of symmetry:

asymmetry Lack of symmetry, the animal cannot be cut along an axis that creates two mirror images.

radial symmetry Has a top, bottom, and a central axis, meaning it can be cut into many equal and identical parts (like a cake or pie).

bilateral symmetry Has a top, bottom, left side, right side, head region, and tail region. A bilaterally symmetrical animal can be divided down the middle into mirror image halves. Humans have this type of symmetry.

Skeletons have one of two types (although not all animals have a skeleton):

exoskeleton Hard skeleton on the outside of body made of a stiff carbohydrate called **chitin**.

endoskeleton Skeleton on the inside of the body made of bones or cartilage.

Digestive systems

no digestive tract Individual cells digest and absorb nutrients from food.

incomplete digestive tract One opening for both food coming in and wastes going out, digestion and nutrient absorption occurs in a single body cavity.

complete digestive tract Two openings, one for food coming in and another for removal of waste. This is the most specialized system—there are separate places for digestion and absorption.

2. Use the following dichotomous key to classify "unknown" animals into various phyla based on their symmetry, skeleton, and digestive system. After you have identified an animal's phylum, fill in the appropriate row in Table 3.7.

Key to Selected Phyla of the Kingdom Animalia

1a. Asymmetry . . . . Phylum Porifera

1b. Radial or bilateral symmetry 2

2a. Radial symmetry 3

2b. Bilateral symmetry 4

3a. Incomplete digestive tract . . . . Phylum Cnidaria

3b. Complete digestive tract. . . . . Phylum Echinodermata

4a. Flattened body with no appendages. . . . . . Phylum Platyhelminthes

4b. Other body types 5

5a. Long, wormlike body 6

5b. Other body types. 7

6a. Unsegmented worms . . . . Phylum Nematoda

6b. Segmented worms . . . . Phylum Annelida

7a. Organism has a shell . . . . Phylum Mollusca

7b. No shell. 8

8a. Organism has hard exoskeleton . . . . Phylum Arthropoda

8b. Organism has endoskeleton. . . . . Phylum Chordata

TABLE 3.7

Unknown Animal	Phylum
A.	
B.	
C.	
D.	
E.	
F.	
G.	
H.	
I.	

LAB EXERCISE 3.5

A Survey of the Kingdom Fungi

Fungi are heterotrophic, like animals, but the way they acquire their food is fundamentally different. Instead of ingesting their food, fungi excrete digestive enzymes and then absorb the predigested food through their cell walls which are made of the protein **chitin**.

Fungi disperse via **spores**, which are single cells with tough outer walls. There are four different phyla of fungi, each distinguished by either a unique method of producing spores or a unique process of sexual reproduction. The classic toadstool-like mushrooms belong to one phylum of fungi, and only one other phylum produces large mushroom structures (although this phylum also encompasses yeast, which does not produce mushrooms at all). Mushrooms are only the spore-producing organs of the fungus. Most of the fungal body is made up of microscopic threads, called **hyphae**, which are found throughout the material the fungus is consuming. Other phyla of fungi include the molds and mildews.

A. Examine various specimens of the Kingdom Fungi.

1. Examine under the microscope the slide of bread mold prepared by your laboratory instructor. Draw and identify the hyphae and the spore-forming structure.

2. Look at the live and dried mushrooms, as well as the slide of a mushroom cap. Sketch the mushroom cap here. Where are the spores found on this structure?

3. Yeasts differ from other fungi because they are not made up of hyphae, but are single-celled structures. Baker's yeast (*Saccharomyces cerevisiae* as well as other species of *Saccharomyces*) is one of the most economically important fungi because of its role in bread making (the carbon dioxide yeast give off as they consume the sugar in an unbaked loaf of bread causes the bread to rise), and in alcohol production (ethanol is a waste product of yeast metabolism in a low-oxygen environment). Examine the yeast and other economically important fungi available in lab.

4. Lichens are a symbiotic association of fungi and algae or photosynthetic bacteria in which each partner benefits from the other. The photosynthesizers receive a protected place to grow and access to water absorbed by the fungus and the fungus receives excess carbohydrates from the photosynthesizers. Observe examples of lichen available in lab.

5. Discuss the following questions and be prepared to share your answers with the lab instructor and your classmates.

   a. The hyphae of fungi are very diffuse and thread-like. Considering how a fungus acquires food from its environment, what advantages does this body form have over a more compact body form?

   b. Based on your observations of the bread mold, what do you think gives bread mold its "fuzzy" appearance?

   c. Some mushrooms are poisonous to humans and some are quite tasty. Considering that the "purpose" of a mushroom is to disperse spores, why might mushrooms have these different traits?

   d. In what environments would you expect lichens to be successful and why?

B. Determine the requirements for fungal growth.

   You have probably had enough experience with moldy food to have formed some hypotheses about which environmental conditions favor or restrict the growth of mold. This exercise will allow you to test one or more of these hypotheses experimentally.

   Rhizopus stolonifera ("Black Bread Mold") is a common mold in the environment and one of the leading causes of food spoilage.

   1. What environmental conditions do you think favor the growth of mold? What conditions do you think restrict its growth? Be creative, and consider the variety of techniques we typically use in food storage to prevent spoilage. Record your ideas in Table 3.8.

TABLE 3.8

Conditions That Favor Mold Growth	Conditions That Restrict Mold Growth

   2. Choose one environmental factor from each column and design a simple hypothesis test to evaluate your hypothesis. You will compare mold growth on a quarter slice of bread that is exposed to the environmental factor in question to growth on a quarter slice of bread that is protected from the same environmental factor. All bread will be inoculated with a small amount of *Rhizopus stolonifera*.

Hypothesis 1:
The following environmental factor - _____ - favors the growth of mold.

Hypothesis 2:
The following environmental factor - _____ - restricts the growth of mold.

Prediction 1: If Hypothesis 1 is correct, I predict that mold growth on the treated bread will be _____ mold growth on the untreated bread after one week.

Prediction 2: If Hypothesis 2 is correct, I predict that mold growth on the treated bread will be _____ mold growth on the untreated bread after one week.

3. With the materials available in the lab, set up your hypotheses tests by applying the appropriate treatments and inoculating each bread slice with a small amount of *R. stolonifera* from the active culture in lab. Be sure to label your petri dishes with your name and the treatment you have applied. Briefly describe your treatments here. Petri dishes will be stored at room temperature for several days and then transferred to a cold place for the remainder of the week.

4. Record your results in Table 3.9. Indicate relative mold growth with the following scale:

   0 = no mold growth, + = small amount of growth (<25% of bread surface), ++ = moderate amount of growth (~50% of bread surface), +++ = bread surface completely covered with mold.

TABLE 3.9

Environmental Factor	Growth on Treated Bread	Growth on Untreated Bread

5. Did your results support your hypothesis? If not, why do you suppose?

6. Did this exercise provide you with any practical information about protecting foods from spoilage? If so, what?

LAB EXERCISE 3.6

A Survey of the Kingdom Plantae

A. Survey the diversity of the plant kingdom.

   The plant kingdom is divided into 10 phyla; of these, only four make up the majority of land plants. These four phyla can be arranged in a timeline that helps to illustrate the major advances in plant evolution.

   Phylum Bryophyta – mosses Early land plants that produce spores for reproduction and do not contain vascular tissue, which in other plants transports water and nutrients throughout the plant body. These plants are thus necessarily small and close to the ground.

   Phylum Pteridophyta – ferns and relatives Some species resemble early land plants as well. Produce spores for reproduction and contain vascular tissue.

   Phylum Coniferophyta – nonflowering seed plants, including conifers Contain vascular tissue and produce seeds, which contain embryos encased in nutritive tissue and a seed coat produced by the parent plant tissue. Sperm is generally transferred inside of resistant pollen grains.

   Phylum Anthophyta – flowering plants Contain vascular tissue, including cells that increase efficiency of water movement. Flowers lead to specificity in pollen transfer and result in the production of fruit, which serves as a dispersal mechanism for seeds.

   1. Observe the various plant specimens available in the lab or in the field. As you examine the plants, try to determine which of the four major phyla each species belongs in (or which one it is closest to).

   2. Based on your observations in lab and your previous knowledge, which of the four phyla do you think is most diverse? Which is least diverse? Which is most abundant? Which is least abundant?

B. Dissect flowers and fruits.

   People are often surprised to learn that many of the plants we consider "vegetables" are actually fruits from a botanical perspective. A fruit is essentially a mature flower. Generally, fruits can be identified as such by the presence of seeds (although some of the fruits we eat have been artificially selected to be seedless, such as many varieties of oranges, bananas, and grapes). In this exercise, you will dissect a "typical" flower and relate the structures of the flower to various fruits.

   1. Determine a flower's structure. Use Figure 3.2 as a guideline as you dissect the flower(s) available in lab. Try to identify all of the major parts labeled on the diagram on your flower.

   2. Determine the relationship of flower to fruit. A fruit is most typically the ripened ovary of a flower, as described in Figure 3.3. Slice the ovary of the flower you are dissecting horizontally to observe the

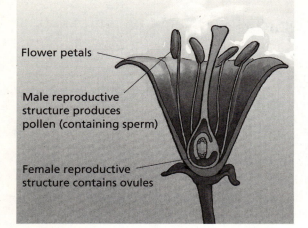

FIGURE 3.2

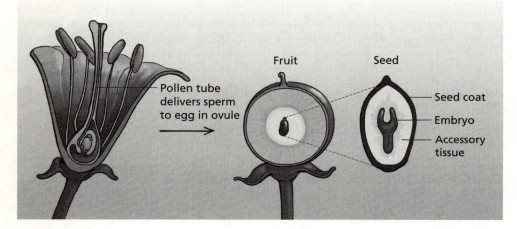

FIGURE 3.3

ovules inside. These structures develop into seeds as the ovary develops into a dispersal structure.

3. Dissect a fruit and identify seeds and other flower parts. Obtain a variety of fruits and examine them for the presence of seeds and any other structures that indicate their relationship to their source flower (for example, petals or stamens still attached). Be prepared to share your dissections with the rest of the class.

C. Determine requirements for photosynthesis.

The one unifying theme for nearly all plants is photosynthesis. Plants use photosynthesis as a mode of nutrition. This means that they use energy from the sun along with water and carbon dioxide gas to produce "food," water, and oxygen gas.

This simple protocol elegantly demonstrates the process of photosynthesis. In it, we will use phenol red, a chemical indicator for carbon dioxide, or CO_2. Phenol red turns yellow in the presence of CO_2 and is pink in its absence.

1. Find the four clean test tubes at your lab bench. If the tubes are not clean, rinse them thoroughly with tap water. Label the tubes 1 through 4.

2. Fill each tube 1/2 full with tap water. Add 10 drops of phenol red to each tube and mix, using a drinking straw as a stirrer. The water should be noticeably pink—if it is not, add more phenol red.

3. Using the straw, carefully blow into tubes 1, 2, and 3 until the phenol red solution *just* turns yellow. (When you blow into the solution, you are dissolving CO_2 from your lungs in the water and the indicator turns from pink to yellow.)

4. Cut two 3-inch long pieces of Elodea, an aquatic flowering plant, from the center table. Place a piece in each of tubes 1 and 2. The Elodea should be completely covered by the water. Place a cork in the top of each tube.

5. Cover tube 2 with a piece of aluminum foil. Be sure the foil completely blocks out all light.

6. If you can, place the tubes in the tube holder in direct sunlight. If there is no direct sunlight, place a bright incandescent light near the sides of the tubes.

7. After 45 minutes, observe and record any color changes in Table 3.10.

TABLE 3.10

	Color at Beginning	Plant	Light Conditions	Results
Tube 1	yellow	Elodea	light	
Tube 2	yellow	Elodea	dark	
Tube 3	yellow	no plant	light	
Tube 4	pink	no plant	light	

8. Discuss the following questions with your lab partners. Be prepared to share your answers with your lab instructor and the class:

1. In which tube(s) has photosynthesis occurred? How can you tell?

2. What is the purpose of tube 3 in the experiment?

3. What is the purpose of tube 4 in the experiment?

Name:_____

Section: _____

TOPIC 3

_____

POST-LABORATORY QUIZ

BIODIVERSITY

1. Describe the currently most accepted hypothesis of the evolutionary relationships among the three domains of life.

2. Most members of the domain Bacteria can be classified into one of two different types based on _____.

3. Name the four kingdoms within the domain Eukarya.

4. What physical characteristic unites all members of the domain Eukarya?

5. Protists are unique among Eukaryotic kingdoms in that members of the kingdom do not share a single, typical, mode of nutrition. What characteristic appears to unite the members of this very diverse kingdom?

6. Describe the typical body form of a member of the kingdom Fungi and explain how this form relates to the fungus life style.

7. The tough cells that fungi produce for dispersal are known as _____.

8. Describe two general characteristics that help define which phyla an animal species belongs to.

9. For each pair, circle the trait that evolved first in the history of plant evolution.

    spores or seeds

    seeds or fruits

    flowers or vascular tissue

    photosynthesis or seeds

10. Describe the relationship between flowers and fruits in flowering plants.

TOPIC 4

Genetics and Inheritance

Learning Objectives

1. Describe how the process of meiosis relates to genetic inheritance.
2. Be able to relate an organism's genetic makeup to its physical appearance.
3. Define genotype, phenotype, homozygous, heterozygous, homologous pairs, alleles, meiosis, gametes, and fertilization.
4. Explain how Mendel's laws can be used to predict the outcomes of matings.
5. Understand that there are situations when Mendel's laws cannot predict the outcome of a mating.
6. Model the processes of meiosis and fertilization in an imaginary organism.
7. Be able to identify whether a trait is recessive, dominant, or codominant.

Pre-laboratory Reading

All of us can think of traits that have been passed down in our own families. You might have hair color similar to your father, or eye shape similar to your mother; you also might worry that you have inherited the trait for an illness that is common among your relatives, such as heart disease or breast cancer. The study of the inheritance of traits is called **genetics**.

Traits are passed from one generation to the next when genes are passed from parents to offspring. Traits are passed on structures called chromosomes. Chromosomes are composed of a molecule called **deoxyribonucleic acid (DNA)** and a variety of proteins. Sections of DNA that encode traits are called **genes**. Each chromosome carries hundreds of genes. Figure 4.1 shows all of the chromosomes found in a human body cell. This picture is called a **karyotype**.

You can see from the karyotype that it is possible to arrange chromosomes into pairs. These pairs are called **homologous pairs**. Humans have 22 homologous pairs of chromosomes and one pair of **sex chromosomes**. Females have two X sex chromosomes and males have one X and one Y sex chromosome.

Homologous pairs of chromosomes carry the same genes, but can carry different versions of each gene. Different versions of a gene are said to be **alleles** of a particular gene. Two individuals with different alleles of a gene will have different appearances or **phenotypes**. This is because an individual's phenotype is determined by the complement of alleles present, called the **genotype**.

Organisms pass their genes to their offspring via cells that are produced by a process of cell division called **meiosis**. In humans, meiosis occurs in the cells of the ovaries and testes to produce egg cells and sperm cells. Cells produced by meiosis are called **gametes** and they contain 1/2 of an individual's genes and chromosomes. Gametes are united at **fertilization**. When a gamete from a female fuses with a gamete from a male, genetic information from each parent will be present in the offspring.

Some genes that are passed on chromosomes are dominant, or expressed when there is one allele present, and some genes are recessive or only expressed when two of the same alleles are present. An understanding of dominant and recessive interactions allows one to predict the outcomes of various crosses. In this lab, you will use imaginary organisms to help you understand the basic principles of genetics.

Autosomes (22 pairs)   **Sex chromosomes** (1 pair)

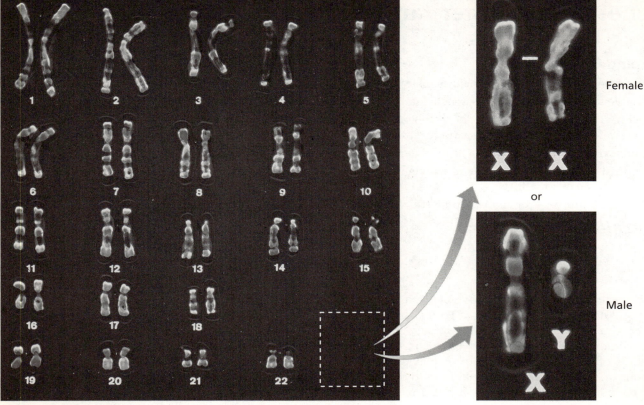

FIGURE 4.1   (CNRI/SPL/Photo Researchers, Inc.)

LAB EXERCISE 4.1

From Genotype to Phenotype

A. Model chromosomes and phenotypes.

Examine the pile of pipe cleaner chromosomes on your lab table. As in a typical cell, these chromosomes are paired; that is, there are two of each type of chromosome. These pairs of chromosomes are homologous to each other. In this case, the members of a homologous pair are the same color. Note that all chromosomes are labeled with one or two letters. These letters are meant to represent a gene or genes that are located on that chromosome.

Because homologous pairs of chromosomes carry the same genes, both of the chromosomes in the pair should be labeled with the same letter or letters, although one may be uppercase and one lowercase. Uppercase and lowercase letters of a given gene represent different alleles of that gene.

Because genetic crosses using live organisms can require many weeks or months to produce offspring, we are going to use a model of a living organism to help illustrate the relationship between the information carried on chromosomes and the characteristics of an organism. Our model is Mr. Potato Head.

   The chromosomes on your lab bench represent all of the chromosomes of Mr. Potato Head. Therefore, this assortment of chromosomes is what you would expect to find in one of his body cells. The letters on the chromosomes represent different alleles for various genetically determined traits. We are going to use the information in these genes to create our version of Mr. Potato Head.

1. The alleles for each gene on the chromosome on your table tell you what your Mr. Potato Head's genotype, or combination of alleles, is. Write his genotype here:

   Tt, _____, _____, _____, _____, _____, _____

2. An individual is said to be **homozygous** when he or she has two of the same alleles of a given gene and **heterozygous** when he or she has two different alleles of a given gene. Is Mr. Potato Head homozygous or heterozygous for each of the gene pairs in the preceding list?

3. Table 4.1 lists the relationship between the genotype of Mr. Potato Head and his phenotype. Use the Mr. Potato Head models at your lab bench to create the phenotype specified by the genotype present on your group's assortment of chromosomes.

TABLE 4.1

Genotype for Each Gene Pair	Phenotype
TT or Tt	Smiling
tt	Sticking out tongue
DD or Dd	Nose
dd	No nose
EE or Ee	Two ears
ee	No ears
MM	Eyebrows and mustache
Mm	Mustache only
mm	No facial hair
QQ	Hat only
Qq	Hat and shoes
qq	Shoes only
AA or Aa	Two arms
aa	No arms
LL or Ll	Eyes, but needs glasses
ll	Eyes, does not need glasses

4. Compare your Mr. Potato Head with others in class. You all started with the same genetic information in your hypothetical nondividing cell; therefore, all of the Potato Heads should have the same appearance. Do they?

LAB EXERCISE 4.2

From Parent to Offspring

Parents pass genes to their children by placing copies of their chromosomes into their gametes (eggs or sperm). Each normal gamete contains exactly half of the genetic information carried by the parent—thus each offspring has an equal number of genes from each parent. However, the half of each parent's genetic material that is passed on is not random. Each gamete contains one copy of every gene found in a human. In other words, each gamete contains one member of each homologous pair of chromosomes.

Recall that the process of cell division that produces gametes containing one homologue of each chromosome is called meiosis.

A. Model meiosis.

You will now model the parent cell represented by the chromosomes on your table undergoing meiosis to produce gametes. Use these instructions and Figure 4.2 as a guide.

1. Prior to meiosis, each chromosome must be replicated. This takes place during the portion of the cell cycle called **interphase**. Prior to this time, the chromosomes are stretched out and difficult to see under the microscope. To model replication, you should do the following:

- Pick up from the center table a matching chromosome for each of the 12 chromosomes in the parent cell you started with.
- Match each original chromosome with its replicate. These replicas are identical copies—each should carry the same allele. These identical copies are called **sister chromatids**. Note that this is really not the way that chromosome copies are produced—there are not extra chromosomes lying around in the cell. Instead, each chromosome is copied in place, and the two copies are held together at a structure called the **centromere**, which is analogous to a twist tie.
- Clip the two copies together with the twist tie centromere. Now you have a representation of the chromosomes in Mr. Potato Head after they have replicated.

2. After interphase, the replicated chromosomes move into a phase of meiosis called **prophase I**. It is during prophase I that homologous pairs of chromosomes can undergo crossing-over. During the next step of meiosis, the homologous pairs of chromosomes move to the center of the cell. This occurs at the stage called **metaphase I**.

- Line up the chromosomes as they appear at this stage of meiosis. Each chromosome is paired with its homologue and members of a pair sit on opposite sides of an imaginary line that runs through the center of the cell, called the equator. (Note that the arrangement of each pair on this line is independent of the arrangement of the other pair. In other words, not all chromosomes carrying lowercase alleles have to be on one side of the cell.) This is called **random alignment** of the homologues.

3. At this point, the homologues separate and the parent cell divides into two daughter cells, each with only one type of each chromosome, although these chromosomes are still attached to their replicate. In Figure 4.2, these steps are labeled **anaphase I** and **telophase I**.

   • Separate each pair of homologues —place one homologue on one side of the table, the other homologue on the other side. Once separated, each cell begins another round of meiosis beginning with **prophase II**.

4. During **metaphase II** of meiosis, the chromosomes in each cell line up at the center of the cell. Model this for each daughter cell on different halves of your lab table. Unlike the first division, they are not paired, but form a single line across the equator.

5. The final division separates the identical copies of the chromosomes. This is illustrated in Figure 4.2 as **anaphase II** and **telophase II**. To model this, you will have to remove the centromere from each pair and place one copy in each of the four resulting daughter cells, now called gametes.

6. Check with your lab mates and/or instructor to determine whether you modeled meiosis correctly. Does each gamete have one of each type (that is, one of each color) of chromosome?

7. Every lab group started with the same chromosomes. Does everyone have identical gametes after meiosis? Why or why not?

8. How many genetically distinct kinds of gametes can be produced when six chromosomes each carry one gene with two different alleles?

B. View meiosis.

1. View the prepared slides showing meiosis. What type of tissue is fixed to the slide?

2. Why do you think this tissue type was used to display meiosis?

3. Try to identify a cell in each stage of meiosis.

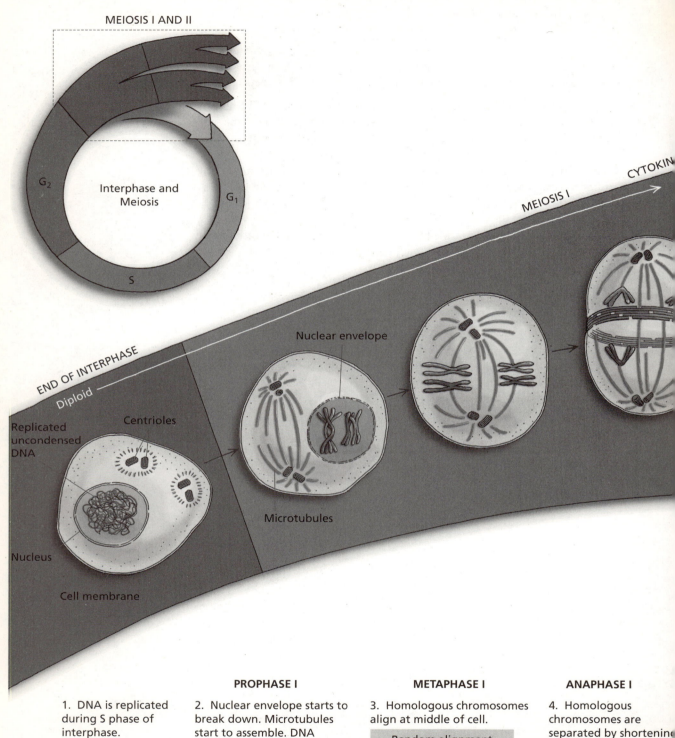

MEIOSIS I AND II

G₂

Interphase and
Meiosis

G₁

S

CYTOKIN

MEIOSIS I

Nuclear envelope

END OF INTERPHASE

Diploid

Replicated
uncondensed
DNA

Centrioles

Microtubules

Nucleus

Cell membrane

PROPHASE I

METAPHASE I

ANAPHASE I

1. DNA is replicated
during S phase of
interphase.

2. Nuclear envelope starts to
break down. Microtubules
start to assemble. DNA
condenses into chromosomes.

Crossing over may occur

3. Homologous chromosomes
align at middle of cell.

Random alignment

4. Homologous
chromosomes are
separated by shortening
of microtubules.

FIGURE 4.2

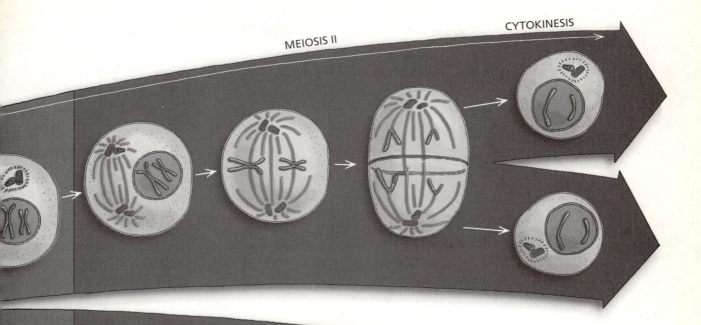

CYTOKINESIS

MEIOSIS II

OPHASE I and
YTOKINESIS

Cytokinesis
ults in two
ghter cells.
clear envelopes
orm.

PROPHASE II

6. Microtubules
lengthen.

METAPHASE II

7. Chromosomes
align at middle
of cell.

ANAPHASE II

8. Sister chromatids are
separated by shortening
of microtubules.

**TELOPHASE II and
CYTOKINESIS**

9. Four haploid
daughter cells result.
Nuclear envelopes
reform.

LAB EXERCISE 4.3

Making Babies

Now our Mr. Potato Heads are going to mate. The reproductive biology of Mr. Potato Head is still shrouded in mystery. Suffice it to say that they are able to mate among themselves, even though there appears to be only one gender.

A. Model fertilization.

1. Swap Mr. Potato Head gametes with a group of students at another lab table. Give them one of your gametes in exchange for one of theirs. Combine this new gamete with one of your remaining three gametes. This models the process of fertilization. The resulting cell should contain six homologous pairs of chromosomes. This offspring will be referred to as offspring A.

2. Repeat the process of fertilization with your remaining two gametes with a group of students at another table to produce offspring B.

3. List the genotypes of the offspring:

Offspring A:

Offspring B:

4. List the phenotypes of the offspring in Tables 4.2 and 4.3. Consult Table 4.1 earlier in the chapter:

TABLE 4.2 **Offspring A**

Mouth type	
Nose	
Ears	
Facial hair	
Clothing	
Arms	
Vision	

TABLE 4.3 **Offspring B**

Mouth type	
Nose	
Ears	
Facial hair	
Clothing	
Arms	
Vision	

5. You will now produce grandchildren Potato Heads. Choose one of these offspring (A or B) to undergo meiosis again. Combine two of your four gametes with gametes from other groups. The offspring produced will be called offspring Y and offspring Z.

6. List the genotypes of these offspring:

Offspring Y:

Offspring Z:

7. List the phenotypes of these offspring in Tables 4.4 and 4.5.

TABLE 4.4 **Offspring Y**

Mouth type	
Nose	
Ears	
Facial hair	
Clothing	
Arms	
Vision	

TABLE 4.5 **Offspring Z**

Mouth type	
Nose	
Ears	
Facial hair	
Clothing	
Arms	
Vision	

8. Build Mr. Potato Heads representing both offspring Y and Z. Compare them with those your classmates made. The initial group of Potato Heads you made were all the same, yet they were able to produce many different-appearing offspring. Potato Heads have only one or two genes on each of 6 pairs of chromosomes. Humans have hundreds of genes on each of 23 pairs of chromosomes. Imagine the number of different offspring possible when humans mate. This explains why siblings can look very different from each other.

LAB EXERCISE 4.4

Confirming Mendel's Ratios

Gregor Mendel, the founder of classical genetics, determined the nature of genes while studying inheritance in pea plants. Through careful **crossing** (mating) of the plants, Mendel noticed that traits appeared in offspring in consistent ratios. Mendel used the term **dominant** to refer to a trait that appeared whenever an individual had the gene for that trait, and **recessive** to refer to a trait that only appeared when the gene for the dominant trait was absent. When he crossed pea plants that were completely dominant against those that were completely recessive, he discovered that all of the offspring displayed the dominant trait. When he crossed these individuals, he found that their offspring displayed a ratio of three dominant for every one recessive. This result led him to conclude that individuals carry two copies of each gene and pass one copy of each gene to each of their offspring. (Actually, Mendel did not use the word gene—that is a more recent term).

A. Determine Mendelian expectations in crosses involving one gene.

1. Use the data from the whole class for offspring A and B to fill in Table 4.6.
2. Did this trait conform to Mendelian expectations?

TABLE 4.6

Group #	# Smiling	# Sticking Tongue Out
1		
2		
3		
4		
5		
6		
Total Number		
Ratio		

B. Determine Mendelian expectations in crosses involving two genes.

When Mendel observed the inheritance of two different characters, he discovered that, at least for the traits he examined, genes are inherited independently of each other. In other words, two traits that occurred together in a plant (such as tall height and white flowers) might not be found together in the second generation of a cross between two different plants. When he examined the ratios of these crosses, he discovered that among the second-generation offspring, for every one individual displaying both recessive traits, there were nine individuals displaying both dominant traits, three displaying the recessive trait for one character and the dominant trait for the other, and three displaying the converse.

1. Use the data from offspring A and B to fill in Table 4.7.

TABLE 4.7

Group #	Two Ears, Nose	Two Ears, No Nose	No Ears, Nose	No Ears, No Nose
1				
2				
3				
4				
5				
6				
Total				
Ratio				

2. Does the ratio conform to Mendel's expected 9:3:3:1 pattern? If it doesn't, why not?

C. Examine the phenotypic ratio for another pair of genes.

1. Use offspring Y and X to fill in Table 4.8.

TABLE 4.8

Group #	Nose, Glasses	Nose, No Glasses	No Nose, Glasses	No nose, No Glasses
1				
2				
3				
4				
5				
6				
Total				
Ratio				

2. Do the phenotypes for these genes conform to the expected 9:3:3:1 ratio? Why or why not?

3. Look closely at the ratios produced for genes D and L. These two genes are linked to each other; that is, they are found on the same chromosome. Do genes that are linked to each other assort independently? Why or why not?

4. Look at the genotype/phenotype relationship for gene Q. Is one allele dominant to the other?

   When both alleles of a gene are expressed (versus one allele being dominant over the other), we say that the alleles of a gene are **codominant** to each other.

5. Look at the genotype/phenotype relationship for gene M. Is one allele dominant to the other? Explain.

   Instead of simulating meiosis and fertilization using pipe cleaners, scientists make use of a tool called a Punnett square. By listing the gametes produced by the male parent on one axis and the gametes produced by the female parent on the other axis, and then combining the gametes to produce offspring in the middle of the Punnett square, scientists can make predictions about the kinds of offspring two parents can produce together.

6. Represent the cross between your offspring Z and your neighbor's for gene E with a Punnett square.

7. Make a second Punnett square representing the cross between your offspring Z and your neighbor's for genes D and T.

TOPIC 4

_____

POST-LABORATORY QUIZ

GENETICS AND INHERITANCE

1. Define the terms genotype and phenotype.

2. Are the following genotypes homozygous or heterozygous?

   A. AA
   B. Aa
   C. aa

3. Draw a homologous pair of chromosomes before replication. Depict each member of the pair carrying different alleles of gene A.

4. Draw two homologous pairs of chromosomes before replication. Draw one homologous pair as heterozygous for gene A and the other as homozygous for gene B. Label the sister chromatids, centromeres, and homologous pairs.

5. What kinds of gametes can be produced by a parent with the genotype WWXXYyZZ?

6. In humans, the gene for tongue rolling (R) is dominant over its allele for nonrolling (r). A man, homozygous for tongue rolling, marries a woman who cannot roll her tongue. Predict the phenotypes and genotypes of any offspring produced from this mating.

7. In peas, tall (TT) is dominant over short (tt). Three experiments in cross-pollination gave the following results. In each case, give the *most probable* genotype for each parent.

   A. tall x tall produced 95 tall, 29 short

B. tall x short produced 50 tall, 0 short

C. tall x tall produced 75 tall, 0 short

8. In peas, the gene for smooth seed coat (S) is dominant to the one for wrinkled seeds (s). What would be the genotypic and phenotypic results of the following matings:

A. heterozygous smooth X heterozygous smooth?

B. heterozygous smooth X wrinkled?

C. wrinkled X wrinkled?

9. Predict the phenotypic ratios from the following cross AaBb x AaBb.

10. Can you design a cross involving two genes that would not lead to the ratio you predicted in the previous question?

TOPIC 5
Mitosis and the Cell Cycle

Learning Objectives

1. Describe the cell cycle.
2. Illustrate the chromosomal condition and significance of each stage of mitosis.
3. Examine the rate of cell division in various tissues.
4. List several ways in which normal cell division is regulated.
5. Explain how mutations to cell cycle control genes can result in tumors.
6. Describe how scientists determine the cancer risk associated with various chemicals and activities.

Pre-laboratory Reading

The process by which cells increase in number by making copies of themselves is called **cell division**. In all **eukaryotic cells** (cells containing a nucleus), the stages of cell division are nearly identical and are collectively known as **mitosis**. Mitosis is one part of the **cell cycle**—a process that includes stages known as **interphase** and **cytokinesis**. The relationship among these phases is diagrammed in Figure 5.1.

As you can see in Figure 5.1, the genetic material inside the cell is duplicated during interphase. Mitosis results in the equal division of this DNA into two daughter cells, and cytokinesis cleaves the single cell into two.

DNA in eukaryotic cells is found in linear structures known as **chromosomes**. A typical human chromosome carries hundreds of genes along its length and is found in a thin, tangled, thread-like form during interphase. Most human cells carry two copies of each chromosome—these pairs are not identical, but they are equivalent, with each member of the pair carrying the same genes.

During the first stage of mitosis, known as **prophase**, chromosomes condense into tightly wound, compact forms that are easily visualized under a light microscope. At the end of prophase, the membrane surrounding the nucleus has disappeared and we can easily see individual, already duplicated chromosomes.

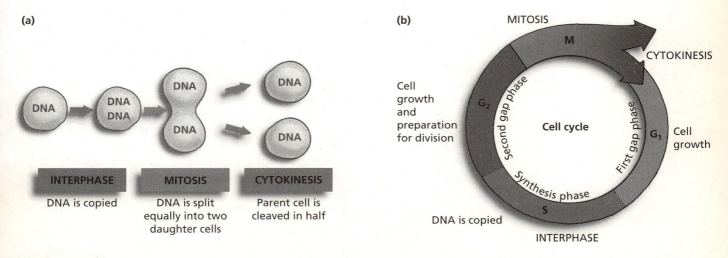

FIGURE 5.1

(b) DNA condensed into chromosomes

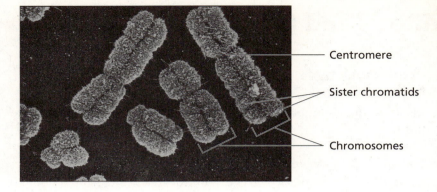

Centromere

Sister chromatids

Chromosomes

FIGURE 5.2   (Biophoto Associates/Photo Researchers, Inc.)

Each chromosome consists of two identical strands called **sister chromatids**, which are attached to each other at a structure called the **centromere** (Figure 5.2).

The second and third stages of mitosis, **metaphase** and **anaphase**, allow for the systematic separation of sister chromatids into two identical daughter cells, each containing a full set of chromosomes. During metaphase, the duplicated chromosomes are aligned across the middle of each cell—the chromosomes are moved to this position by the actions of **microtubules**, protein structures that radiate out from **centrioles** at the two poles of the cell and attach to the centromere of each chromosome. Aligning the chromosomes at the middle of the cell ensures that each daughter cell will contain one copy of each chromosome. The sister chromatids of each chromosome are separated during anaphase as the microtubules contract toward the poles of the cell. After the sister chromatids are separated, they are each referred to as chromosomes.

The final stage of mitosis is known as **telophase**, during which the nuclear membrane reforms and the chromosomes decondense and become diffuse and threadlike once more. The details of cytokinesis vary among the kingdoms of eukaryotes, but result in the formation of two separate cells. The process of mitosis is illustrated in Figure 5.3.

The purpose of cell division in humans is to heal wounds, replace damaged or dead cells, and help tissues and organs grow. Thus, not all cells in the body go through the cell cycle at the same rate and at the same time. Tissues where rapid cell replacement is necessary, such as skin and the lining of the digestive system, will have many cells in various stages of mitosis at any one time. Tissues where cell damage is more rare, such as the liver or heart muscle, will have fewer cells undergoing mitosis. When cell division is working properly, it is tightly controlled; that is, cells are given signals for when, and when not, to divide. Cells that somehow escape this control become **cancerous**, meaning that they are dividing without control. Cancer cells escape cell cycle control because one or more of the genes that affect the process of cell division have been damaged, or **mutated**.

Genes that encode for proteins that regulate the cell cycle are called **proto-oncogenes**. When they are mutated, they are called **oncogenes**—or "cancer genes." Many proto-oncogenes encode for **growth factors** or encode for proteins that respond to the presence of growth factors. A normal growth factor stimulates division only when the cellular environment is favorable and all of the conditions for division have been met. Other proto-oncogenes are **tumor suppressors**, coding for proteins that suppress cell division if conditions are not favorable. These proteins detect and repair DNA damage. Tumor suppressor proteins prevent uncontrolled division when a growth factor or cell division gene mutates and becomes nonfunctional. For cancer to occur, both types of genes in a single cell must be mutated.

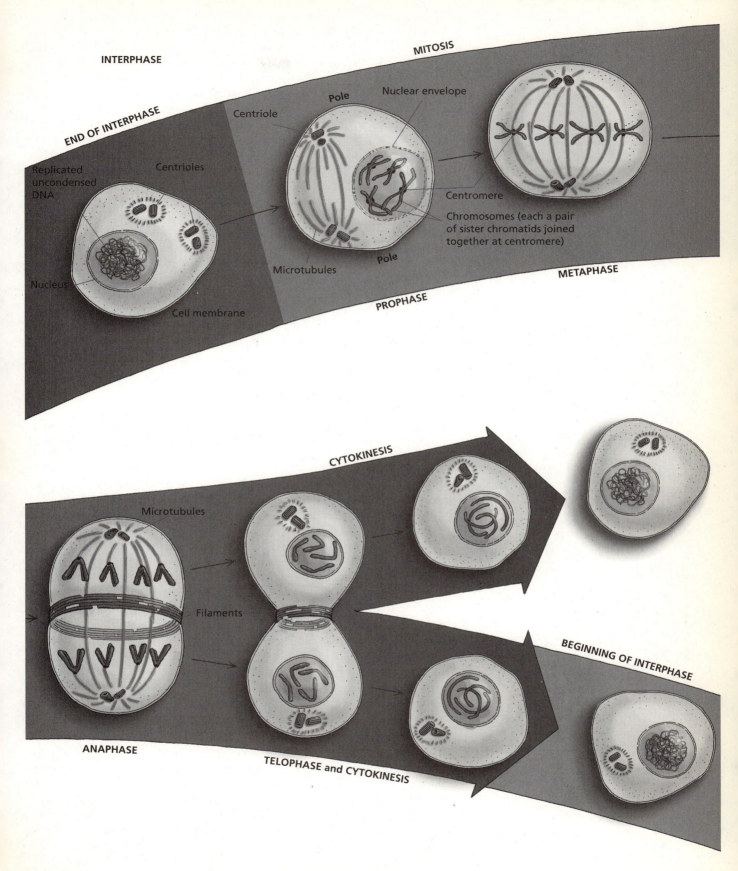

INTERPHASE

MITOSIS

END OF INTERPHASE

Pole

Centriole

Nuclear envelope

Replicated
uncondensed
DNA

Centrioles

Centromere

Chromosomes (each a pair
of sister chromatids joined
together at centromere)

Nucleus

Microtubules

Pole

Cell membrane

PROPHASE

METAPHASE

CYTOKINESIS

Microtubules

BEGINNING OF INTERPHASE

Filaments

ANAPHASE

TELOPHASE and CYTOKINESIS

FIGURE 5.3

After a cell escapes cell cycle control and tumor suppression, it will continue to divide. If the cell is in a relatively immobile tissue, a mass of cells, called a **tumor**, forms. Additional mutations within tumor cells can enable the cancer to spread to other tissues and increase its rate of growth. The excess cells produced by cancer rob other tissues of energy and interfere with the normal functioning of the body.

Cancer-causing mutations to DNA occur most often as a result of the action of DNA-damaging factors called **carcinogens**. There are many known carcinogens, including ultraviolet light, tobacco smoke, and pesticides such as DDT. There are also many suspected carcinogens—chemicals and other environmental factors that can potentially cause DNA damage, but that do not have a clearly established link to human cancers. Carcinogens and suspected carcinogens are also known as cancer **risk factors**. The branch of biology called **epidemiology** investigates the relationship among risk factors and disease. Cancer epidemiologists attempt to determine whether exposure to a particular risk factor is associated with a higher risk of a particular form of cancer, typically by retroactively surveying individuals with that form of cancer, but also by exposing other animals to the risk factor. Epidemiologists also devise strategies for reducing individual exposure to known or suspected carcinogens.

LAB EXERCISE 5.1

Model the Process of Mitosis

This simple exercise is designed to help you visualize the dynamic process of mitosis. Use the illustrations in the pre-laboratory reading to help you walk through this process.

1. Obtain *three pairs* of pipe cleaners. Each pair should be a different color. Both members of the pair should be labeled with the same letters, although these letters do not need to be both uppercase or both lowercase in any pair.

   Each pipe cleaner represents a chromosome, and the labels represent a single gene on the chromosome. This set of chromosomes represents the genetic material found in a cell at the beginning of interphase. Illustrate the entire cell at this stage by placing the pipe cleaners on your lab table, encircling them with twine indicating the nuclear membrane, and encircling the nucleus with yet another strand of twine, indicating the cell membrane. Place two of the centriole spools in the cell, outside the nuclear membrane.

2. You can illustrate the chromosome duplication that occurs during interphase simply, by obtaining a duplicate of each of the pipe cleaners in your original collection. The duplicates should be identical to the original; in other words, the letter indicating the gene on the duplicate should be the same case as the original. "Clip" each pair of chromosomes together with the available centromeres. Replace the duplicated chromosomes in the nucleus.

3. Illustrate prophase of mitosis by removing the twine symbolizing the nuclear membrane, which dissolves at this stage. Move the centrioles to the poles of the cell (top and bottom) and unravel the "microtubules" from each. Wrap a strand of microtubule wire from each centriole to different sister chromatids in each duplicated chromosome. The wire should be as close to the centromere as possible, and both chromatids should be attached to the microtubules at the same side of the centromere (see Figure 5.4).

4. Illustrate metaphase of mitosis by pulling the centrioles toward opposite poles so that their attached chromosomes are lined up at the equator of the cell.

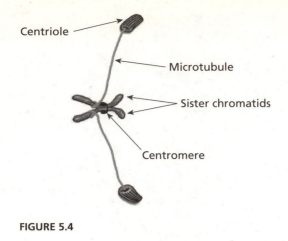

Centriole

Microtubule

Sister chromatids

Centromere

FIGURE 5.4

5. Illustrate anaphase of mitosis by sliding the centromere off each pair of sister chromatids and by winding up the microtubules on each centriole. This should cause each chromatid to move away from its sister, toward opposite poles of the cell.

6. Illustrate telophase of mitosis by pinching the twine of the cell membrane near the equator of the cell, effectively separating the original cell into two compartments, each containing a full set of unduplicated chromosomes. Reform the nuclear membranes around the chromosomes in each cell compartment.

7. Illustrate cytokinesis by clipping the twine of the cell membrane and reforming it into two distinct cells.

LAB EXERCISE 5.2

Hypothesizing About, and Estimating the Rate of, Cell Division

We can estimate the rate of cell division in various tissues by microscopic examination of sections of this tissue stained to highlight the genetic material. Active growth in plants occurs at the tips of stems and roots. A simplified diagram of the growth regions of an onion root tip appears in Figure 5.5. The root cap functions to protect the actively growing regions of the root from the soil, while the apical meristem serves as the source of cells for mitosis.

A. Hypothesize about relative rate of cell division in various regions of an onion root tip.

Using the information provided in Figure 5.5 and the paragraph above, generate a hypothesis of the relative ranking of the rate of cell division in the following regions of an onion root tip and record the information in Table 5.1. The ranking should be from 1 (fastest) to 4 (slowest).

TABLE 5.1

Region of Root Tip	Relative Ranking of Cell Division Speed
Root Cap	
Apical Meristem	
Zone of Elongation	
Zone of Maturation	

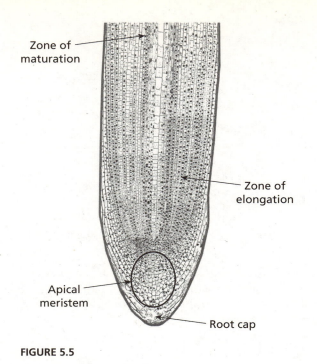

FIGURE 5.5

B. Test your hypothesis by estimating the rate of cell division.

1. Work in groups of four students for this exercise.

2. Each student should examine the onion root tip slide provided under low-power magnification and generally identify the four regions described in the diagram. All of the members of the group should feel comfortable that your identification of these stages is correct. Ask your lab instructor for confirmation if necessary.

3. Assign one region to each student in your group. After you know the region for which you are estimating cell division rate, observe it under high-power magnification. In most regions, you should be able to see the square or rectangular cells and the genetic material within the cell. You might notice the linear chromosomes in various positions in some of the cells. These cells are somewhere in the process of mitosis. However, most cells will have an obvious nucleus with indistinctly stained genetic material, indicating that the chromosomes have not condensed.

4. Count all of the cells in a field of view, noting which are in the process of mitosis, and fill in Table 5.2. Repeat this for a total of three different fields of view for your assigned region. You might need to swap slides with another group member to provide enough different cells to observe.

   Region of root tip examined: _____

5. Each student should enter his or her results in the appropriate row of Table 5.3. The relative rank of division rate is directly related to the percent of mitotic cells in each stage.

6. Compare your results to your hypothesis in part A. Were your predictions accurate? If not, why? Be prepared to share your results and discussion with your lab instructor and classmates.

TABLE 5.2

Number of Cells in Stage	Interphase (Not Dividing)	Mitotic (Dividing)	Total Cells in View
View 1			
View 2			
View 3			
Total (1 + 2 + 3)			
Total Number Counted in Stage/Total Cells Counted			
Percent of Cells in Each Stage (Previous Column × 100)			

TABLE 5.3

Stage	Percent in Interphase	Percent Mitotic	Relative Rank of Division Rate
Root Cap			
Apical Meristem			
Zone of Elongation			
Zone of Maturation			

LAB EXERCISE 5.3

Explore Gene Regulation

You already know that not all genes are **expressed**, or translated into proteins, in all cells. For instance, the gene that codes for the protein that digests milk sugar, the enzyme lactase, is only expressed in cells of the digestive system, but not in eye cells. Similarly, the genes that control cell division are only normally expressed in cells under certain specific conditions. Cancer, or uncontrolled cell division, is ultimately a failure of gene expression. In this exercise, we will examine one model of the regulation of gene expression in living organisms.

Esherischia coli is a bacterial species that is normally found in the digestive system of humans and other animals. *E. coli*'s preferred food source is the sugar glucose, but these bacteria also can digest other sugars in the absence of glucose. One of the alternative sugars *E. coli* can use is lactose. Because *E. coli* does not "prefer" lactose, the protein enzyme that begins the process of lactose digestion (called B-galactosidase in bacteria) is not continually expressed in the cell; this is sensible, because the cell should not use resources to build an enzyme that is unnecessary for its survival *unless* it is truly needed. Therefore, the gene for B-galactosidase is **inducible**; that is, it is typically "turned off," but can be "turned on" under some conditions.

A. Visualize the induction of the B-galactosidase gene.

   B-galactosidase breaks lactose into its two component sugars, glucose and galactose. This enzyme also can break the chemical o-nitrophenol-*B-D*-galactopyranoside (ONPG) into two parts. When ONPG is split by B-galactosidase, it releases the compound o-nitrophenol, which is yellow in color. Thus, we can test for the presence of the enzyme in a bacterial culture by adding ONPG and watching for a color change. ONPG is an **indicator** that tells us if a cell is expressing the B-galactosidase gene.

1. Work with three other students for this exercise. Wear latex gloves and eye protection.
2. Three bacterial cultures are available in the lab labeled A, B, and C. You will perform the same procedure for each culture.
3. Add 1 milliliter of water to a test tube. Label the test tube with the culture designation (A, B, or C).
4. Add 2 drops of chloroform to the test tube. Make sure that the "blob" of chloroform drops to the bottom of the tube rather than resting on the surface of the water (if it rests on the surface, it will quickly evaporate).
5. Add 4 drops of sodium dodecyl sulfate (SDS) to the test tube.
6. Use a toothpick to transfer a small sample of bacteria from the petri dish to the tube. Drop the toothpick into the test tube and hold the tube on the vortexer for 10 seconds. Use a forceps to remove the toothpick and discard.
7. Make sure that the ONPG mixture does not already have a yellow color to it. If it does, it has been contaminated by B-galactosidase—ask your lab instructor for a fresh supply. Add 4 drops of the ONPG mixture to the tube. DO NOT ALLOW THE ONPG DROPPER TO TOUCH THE MOUTH OR SIDES OF THE TUBE.
8. Set the tube in the rack and check on it after 20 to 30 minutes. If B-galactosidase is present, the liquid in the tube will gradually turn yellow.
9. Fill in the results in Table 5.4.

TABLE 5.4

Culture	B-galactosidase Present?
A	
B	
C	

10. Empty the test tubes into the waste bottle (NOT down the drain!). Rinse the tubes with water. Discard the latex gloves. Wash your hands thoroughly with soap and warm water.
11. Discuss the following questions and be prepared to share your answers with your instructor and classmates.

    A. The production of B-galactosidase is induced by the presence of lactose in the environment of the bacteria. Which culture(s) do you think were grown in the presence of lactose? Why?

    B. Some mutant *E. coli* will produce B-galactosidase even without the presence of lactose. Does this information also affect your answer to question 1? Might one of the B-galactosidase positive strains be growing without the presence of lactose, but still expressing the gene? Can you determine this from the test we performed?

B. Relate the regulation of B-galactosidase in *E. coli* to the regulation of cell division in humans.

   The inducible gene for B-galactosidase in *E. coli* is similar, in basic principle, to the inducible genes controlling cell division in our bodies. In the case of B-galactosidase, the substance that turns on the gene is lactose; in the case of cell division control genes, the substance that turns on the gene is known as a growth factor. The following questions ask you to relate what you saw in the case of the *E. coli* strains to what happens in normal and cancerous cells.

   Table 5.5 identifies the cultures used in part A of this exercise. Review this table and answer the following questions. Be prepared to discuss your answers with your lab instructor and/or classmates.

TABLE 5.5

Culture	*E. coli* Strain	Food Source
A	Wild-type (normal)	Nutrient broth
B	Wild-type	Nutrient broth + lactose
C	I-(unable to turn off enzyme production)	Nutrient broth

1. Which of the *E. coli* cultures is equivalent to a normal cell (that is, one without cell-cycle gene mutations) that is not receiving a signal to divide? Explain.

2. Which of the *E. coli* cultures is equivalent to a normal cell that *is* receiving a signal to divide? Explain.

3. Which of the *E. coli* cultures is equivalent to a cancerous cell with a mutant growth factor gene? Explain.

4. Another genetic factor that leads to the development of cancer is a mutation in a tumor suppression gene. When these genes are functional, the proteins they produce inhibit the division of cells with other genetic problems; so even if the cell is producing growth factors when it shouldn't, a functional tumor suppressor will stop the cell from dividing very rapidly. How could the function of a tumor suppressor be modeled in this system?

5. How do mutations in proto-oncogenes lead to cancer?

LAB EXERCISE 5.4

Evaluate the Relationship Between a Risk Factor and Cancer

Nearly 90% of the people who develop cancer do not have an inherited mutation that makes them especially susceptible to cancer; instead, they have *acquired* the crucial mutations throughout their lifetimes. Thus, there is enormous interest in identifying and controlling carcinogens in our environment. Epidemiologists face a difficult challenge when investigating the environmental causes of cancer. Because people are exposed to a diverse, and sometimes interacting, set of environmental conditions over their lifetimes, it can be difficult to demonstrate whether any single agent is carcinogenic. Read the following summary of the evidence for a relationship between second-hand smoke (also known as environmental tobacco smoke, or ETS) and cancer and answer the following set of questions. Be prepared to share your answers with your instructor and classmates.

- Active smoking is the principal cause of lung cancer. Exposure to ETS involves exposure to the same numerous carcinogens and toxic substances that are present in tobacco smoke produced by active smoking.

- More than 50 studies of ETS and lung cancer risk in people who have never smoked, especially spouses of smokers, have been published during the last 25 years. These studies have been carried out in many countries. Most showed an increased risk of lung cancer, especially for persons with higher exposures to ETS. The excess risk increases with increasing exposure.

- Experiments testing the carcinogenicity of ETS are typically performed on rodents. Machines that simulate human active smoking patterns by blowing smoke into the environment produce ETS for the animals. The experimental systems do not exactly simulate human exposures, and the tumors that develop in animals are not completely representative of human cancer.

1. Is this evidence sufficient to prove beyond a reasonable doubt that ETS causes lung cancer in humans? Why or why not?

2. What sort of evidence would be required to establish a cause and effect relationship between ETS and cancer in humans? Is this evidence likely to become available soon?

3. Given your answers to the previous two questions, explain why it is so difficult to prove a link between any environmental factor and an increased risk of cancer in humans.

4. People in industrialized countries are exposed to dozens, if not hundreds, of suspected carcinogens in the course of a year. We are also exposed to many other chemicals whose carcinogenic potential is unknown. And, new chemicals are produced and released into the environment continually. As a society, we potentially have three ways to address the environmental

causes of cancer: 1) a new chemical is innocent until proven guilty—in other words, the chemical can be released widely with little restriction until science provides overwhelming evidence that the chemical causes cancer in the typical levels people are exposed to; 2) a new chemical is innocent until it appears guilty—the chemical can be released widely with little restriction until some evidence appears that it might cause cancer; 3) a new chemical is guilty until shown to be innocent—the chemical is highly restricted in use and release until studies indicate that the chemical is benign. Standards in the United States fall between option 1 and option 2. Do you think this is appropriate? Explain your answer.

5. Cancer has proven to be highly resistant to a cure. There are simply too many types of cancer and too many gene mutations that can lead to cancer to develop a one-size-fits-all cure. Despite many years and billions of dollars of research, cancer is still one of the leading causes of death in industrialized countries. Do you think more effort should be put into research into cancer prevention? Do you know what behaviors increase or decrease your cancer risk?

TOPIC 5
_____

POST-LABORATORY QUIZ

MITOSIS AND THE CELL CYCLE

1. A human cell contains 46 chromosomes. How many chromatids would be found in a cell at the beginning of mitosis?

2. In this cell, the nuclear membrane has just disappeared and the chromosomes become visible as individual structures. Draw the cell at the next stage of mitosis.

3. During what stage of the cell cycle does DNA synthesis occur?

4. Describe what occurs during the anaphase of mitosis.

5. What occurs during the process of cytokinesis after mitosis?

6. Examine the following data in Table 5.6.

TABLE 5.6

Tissue	% Cells in Interphase	% Cells Mitotic
Skin	75	25
Liver	90	10
Cornea	99	1

Which tissue has the greatest rate of cell division?

7. What is the significance of the line of duplicated chromosomes that forms at the cell's equator during metaphase of mitosis?

8. What is cancer?

9. Examine the data in Table 5.7.

TABLE 5.7

Cell Line	Growth Conditions	Behavior of Cells in Culture Dish
A	Nutrients	No mitosis
A	Nutrients + growth factor	Mitotic
B	Nutrients	No mitosis
B	Nutrients + growth factor	No mitosis
C	Nutrients	Mitotic
C	Nutrients + growth factor	Mitotic

Which of the cell lines is likely cancerous?

10. Dioxin, a byproduct of the production of vinyl, is known to cause damage to DNA in cells in culture. How would an epidemiologist determine whether dioxin is a carcinogen?

TOPIC 6

DNA Structure, Synthesis, and Fingerprinting

Learning Objectives

1. Describe the structure of DNA.
2. Understand and model the process of semiconservative DNA replication.
3. Understand how DNA fingerprinting can be used to identify individuals.

Pre-laboratory Reading

The DNA molecule is the molecule of heredity that is passed from parents to offspring. Figure 6.1 shows this molecule's structure.

The DNA molecule is composed of two strands. Each strand has a backbone made up of a sugar, **deoxyribose**, and a phosphate group. The backbone holds the strand together, but does not contain any genetic information. The alignment of the two strands of a DNA molecule is referred to as **antiparallel**. In other words, the backbones are flipped with respect to each other.

In addition to the sugar phosphate backbone, the DNA molecule contains chemicals called **nitrogenous bases**. The nitrogenous bases found in DNA include adenine (A), cytosine (C), guanine (G), and thymine (T). Different DNA molecules have different orders of nitrogenous bases.

The nitrogenous bases are connected to the backbone via the sugar molecule, which in turn, is connected to a phosphate group. When taken together, the nitrogenous base, the sugar, and the phosphate are called a **nucleotide**. Different nucleotides in DNA have different nitrogenous bases connected to the sugar molecule.

The nitrogenous bases of the two strands pair with each other according to the **base pairing rules**: A always bonds with T, and G always bonds with C. Bonds between bases hold the two strands of the molecule together. These **hydrogen bonds** are symbolized by dotted lines because they are weaker than typical chemical bonds and are more easily broken.

It is relatively straightforward for a cell to copy its DNA because the parental DNA molecule can be used as a template for the synthesis of new, so-called daughter strands. The parental strand is unwound and new nucleotides are added according to the base pairing rules.

DNA can be used as a molecular identification tag when subjected to the process of DNA fingerprinting. This is based on the fact that different people will always, unless they are identical twins, have different DNA sequences.

In this laboratory, you will make a three-dimensional model DNA and then manipulate it to understand DNA synthesis and DNA fingerprinting.

(a) DNA double helix is made of two strands.   **(b)** Each strand is a chain of antiparallel nucleotides.

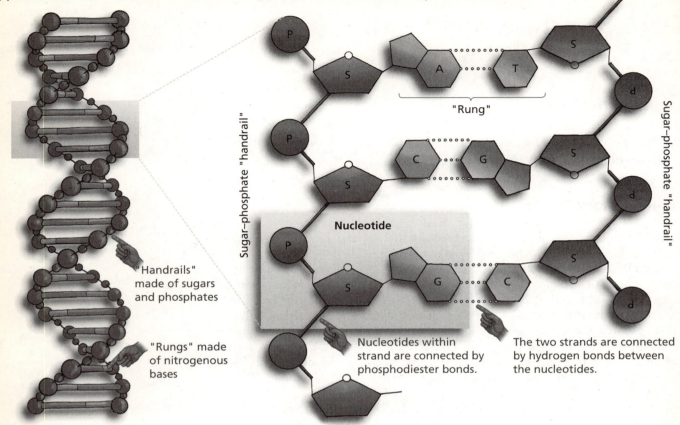

"Handrails" made of sugars and phosphates

"Rungs" made of nitrogenous bases

Sugar–phosphate "handrail"

"Rung"

Nucleotide

Sugar–phosphate "handrail"

Nucleotides within strand are connected by phosphodiester bonds.

The two strands are connected by hydrogen bonds between the nucleotides.

(c) Each nucleotide is composed of a phosphate, a sugar, and a nitrogenous base

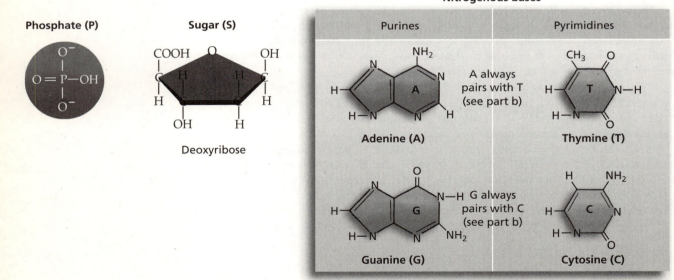

Phosphate (P)

Sugar (S)

Deoxyribose

Nitrogenous bases

Purines	Pyrimidines
Adenine (A)	Thymine (T)
Guanine (G)	Cytosine (C)

A always pairs with T (see part b)

G always pairs with C (see part b)

FIGURE 6.1

LAB EXERCISE 6.1

Constructing a 3-D Model of DNA

A. Building DNA

Obtain a note-card bearing a printed DNA sequence from your lab instructor. This is a list of the nucleotide bases from one strand of a segment of DNA molecule. Note that this is a very short strand—in humans most chromosomes contain around 100 million bases!

1. Note the sequence ID# here: _____

2. Keep track of which nucleotide each colored bead represents in Table 6.1.

TABLE 6.1

NUCLEOTIDE	COLOR
A	
C	
G	
T	

3. Use the twine and beads provided to build a 3-D double-stranded model of this DNA strand. The twine represents the backbone (sugars and phosphates) and the four different colored beads represent the four nitrogenous bases (A, C, G, T).

LAB EXERCISE 6.2

DNA Replication

DNA replication occurs in the nuclei of your cells prior to the process of cell division. In an earlier laboratory, you practiced the process of meiosis, the division that, in humans, results in the production of gametes called sperm and egg cells. The gametes produced by meiosis carry their own unique complement of genetic information.

The type of cell division that produces genetically identical daughter cells and takes place in all of our cells except for the testes or ovaries is the process of **mitosis**. Producing genetically identical copies is fairly straightforward for cells because of the complementary nature of the two DNA strands.

One of the enzymes that facilitates DNA replication functions by first unwinding the DNA molecule. This unwinding occurs as the hydrogen bonds holding nitrogenous bases are broken. Once unwound, each strand of the DNA molecule can be used as a template for the synthesis of a new daughter strand of DNA. The type of enzyme that joins adjacent nucleotides to each other is called a **DNA polymerase**. Figure 6.2 illustrates the process of DNA replication.

1. Use the double-stranded DNA molecule you made in the previous laboratory exercise as a template for the synthesis of two new daughter DNA

(a) DNA replication

(b) The DNA polymerase enzyme facilitates replication

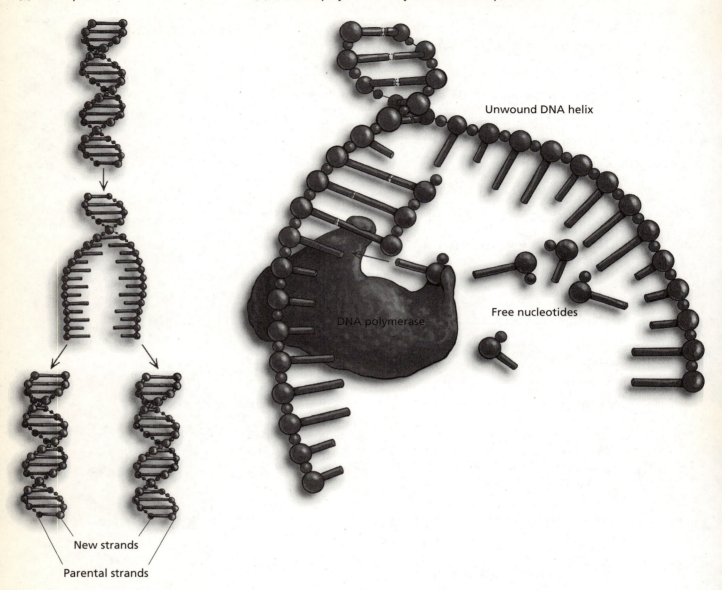

Unwound DNA helix

DNA polymerase

Free nucleotides

New strands

Parental strands

FIGURE 6.2

molecules. Do this by first unwinding the double-stranded DNA and then following the rules of complementarity to make newly synthesized daughter DNA strands. Use a different color of twine when synthesizing the daughter DNA molecules.

2. Describe the daughter DNA molecules you produced in terms of their origin. Are they composed of all daughter DNA, all parental DNA, or some parental DNA and some daughter DNA?

DNA synthesis is sometimes referred to as **semiconservative** replication because each daughter strand is composed of half parental DNA and half newly synthesized daughter DNA. In other words, half of the original parental DNA molecule is conserved in each daughter molecule.

3. Show your instructor the daughter DNA molecules you have made to be certain you made them correctly, and then disassemble one of them.

Note that replication is the point at which changes to the DNA sequence, called **mutations**, can occur. Errors are relatively rare, but the human genome contains over six billion bases—so even with a very low error rate, each round of replication can potentially cause mutations.

LAB EXERCISE 6.3

DNA Fingerprinting

Understanding the basics of DNA structure and replication can help you understand how DNA is manipulated to create DNA fingerprints. In this lab exercise, you will use the DNA sequences you assembled to determine the paternity of a child.

1. Find the sequence ID for your genetic sequence on the first page of this lab. Everyone at your lab table should have a sequence ID with the same first letter.

2. Link the DNA sequences end to end in numerical order according to their sequence ID (for example, M1, M2, M3, M4). This is your table's chromosome. Chromosomes are not this short, of course, and in most species, chromosomes contain a lot of DNA that does not code for genes, but this model will work for our next exercise.

   A DNA fingerprint is a unique pattern of DNA fragments that results when an individual's unique DNA is chopped up by chemicals called restriction enzymes. These enzymes cut DNA at specific sequences. The restriction enzyme we will use in our simulation is *Ham*III, which cuts DNA wherever the sequence GGCC is found (on either strand). Thus, in a DNA molecule:

   the restriction enzyme cuts at the point marked by the vertical line, resulting in three DNA fragments of different lengths.

3. Examine your table's chromosome. Use the scissors provided to cut the DNA molecule at every *Ham*III restriction site.

4. Visually represent the fragments that result as follows:

   a. Count the number of *bases* on each fragment. (Remember that this is the number of bases on a single strand *times two*.)

   b. Find your "lane" on the following table and draw lines in the lane that correspond to the length of each fragment. This is your chromosomes' fingerprint.

5. Visit the other tables to collect their fingerprints and fill in the appropriate lane in Table 6.2.

TABLE 6.2   Chromosome ID

Fragment Size	M	C	W	X	Y	Z	Standard
100							▬▬▬
95							
90							
85							▬▬▬
80							
75							▬▬
70							
65							
60							▬▬▬
55							
50							
45							
40							
35							▬▬▬
30							
25							
20							
15							
10							
5							▬▬▬

What you have just drawn is equivalent to the DNA fingerprint produced by a medical forensics laboratory. In a real DNA fingerprint, the fragments are separated from each other by size according to how quickly they move through a gelatinous substance called a gel. DNA has a slight negative charge, and it will be attracted to a positive charge. To separate the fragments, the chopped-up DNA is placed on one side of the gel and the fragments are then subjected to an electric current. As the fragments migrate through the gel toward the charge, the larger fragments move more slowly than smaller fragments because the gel impedes the progress of the larger ones more than the smaller ones. Over time, the distance between fragments of different sizes grows.

The standard column on the preceding fingerprint corresponds to a standard used in DNA labs—by using fragments of known size on this column, you can estimate the size of fragments in other columns. Because you already know the size of your fragments, using a standard for this demonstration is unnecessary.

The DNA fingerprints you just produced will help you determine the paternity of a child whose fingerprint is in lane C. The mother of this child is known, and her fingerprint is in lane M. Because a child's entire DNA was inherited from its mother and father, any DNA fragment possessed by the child must be present in one of his or her parents. Thus, the father of this child is the one with the fragments that fill in the gaps—the DNA fragments it could not have received from its mother.

6. Which fingerprint—W, X, Y, or Z—belongs to this child's biological father?

TOPIC 6
_____

POST-LABORATORY QUIZ

DNA STRUCTURE, SYNTHESIS, AND FINGERPRINTING

1. If one strand of a DNA molecule has the sequence AGCTTCAGT, the other strand should have the sequence:

2. List the components of a nucleotide. Which of these differ between different nucleotides?

3. Using two differently colored pencils (or a pen and a pencil) diagram a double stranded DNA molecule undergoing replication. Start with two intertwined lines of one color representing the parental DNA molecule. Diagram the results of two rounds of semiconservative DNA replication using the second color to represent the daughter DNA.

4. When does DNA synthesis occur?

5. Why are chromosomes sometimes depicted as linear structures and sometimes as Xs?

6. Why might DNA fingerprinting be more useful in identifying individuals than blood typing analysis?

7. What do we call mistakes in DNA replication that are passed on to offspring?

8. If a DNA fingerprint from a suspect matches blood found at the scene of a crime, should the suspect be convicted?

9. Why might two related individuals share more similar DNA fingerprints than unrelated individuals?

10. What biological molecules act as molecular scissors to cut DNA at specific locations?

Transcription, Translation, and Genetically Modified Organisms

Learning Objectives

1. Understand and model the process of transcription.
2. Understand and model the process of translation.
3. Model the effects of a mutation on protein shape and function.
4. Participate in a debate about genetically modified organisms.

Pre-laboratory Reading

During the last laboratory session, you learned about DNA structure and replication. During today's laboratory session, you will explore the significance of differences in DNA sequences. Sequences of DNA that code for the production of proteins are called **genes**. A gene can be thought of as a set of instructions for the assembly of a protein.

Proteins are produced by the stepwise processes of **transcription** and **translation**.

During the process of transcription (see Figure 7.1), the DNA comprising a gene is used as a template in the production of a molecule called **RNA (ribonucleic acid)**. RNA differs from DNA in that the sugar is **ribose** (not deoxyribose) and instead of thymines, there are **uracils (U)**. The RNA produced by transcription is called **messenger RNA (mRNA)** because it carries the message from the DNA. These mRNA transcripts are single stranded.

The production of the mRNA molecule from the DNA template strand requires the help of an enzyme called **RNA polymerase**. This enzyme ties together adjacent RNA nucleotides as they are being added to the growing mRNA transcript. The transcript is produced when complementary base pairs

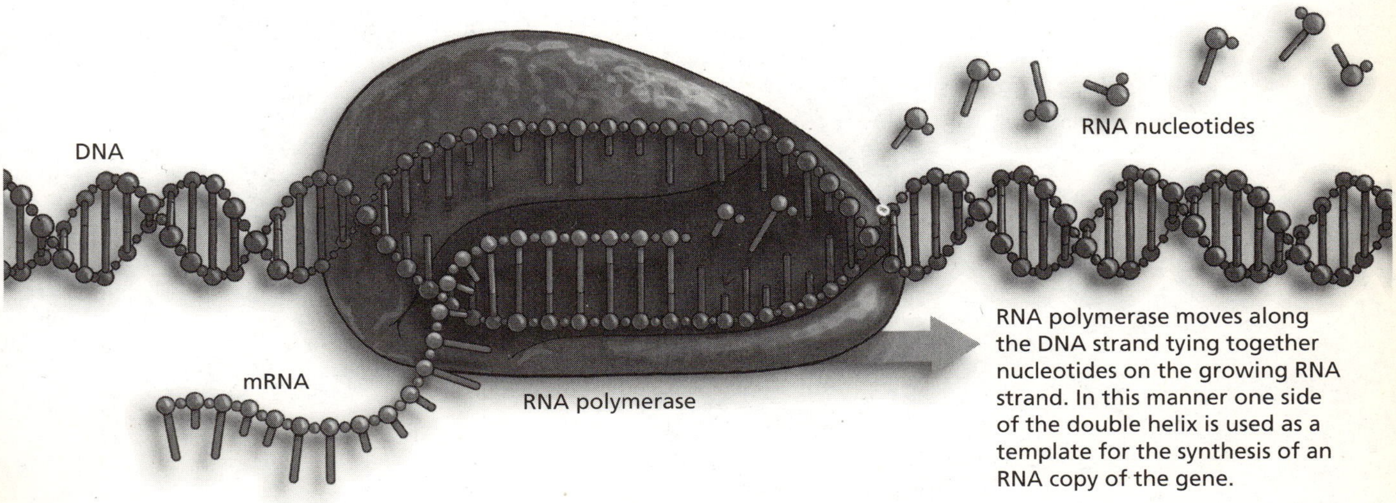

DNA

RNA nucleotides

mRNA

RNA polymerase

RNA polymerase moves along the DNA strand tying together nucleotides on the growing RNA strand. In this manner one side of the double helix is used as a template for the synthesis of an RNA copy of the gene.

FIGURE 7.1

TABLE 7.1   **DNA:RNA Base-Pairing Rules**

DNA	: RNA
C	: G
G	: C
A	: U
T	: A

transiently form with the DNA template strand. RNA makes base pairs with DNA according to the rules listed in Table 7.1.

After the mRNA is produced by transcription, its message is decoded and a protein is produced by the process of translation (see Figure 7.2). Translation occurs in the cytoplasm of cells on structures called **ribosomes**. Ribosomes help anchor the mRNA and help synthesize the protein coded for in the DNA. The mRNA is "read" or deciphered by the ribosome as a series of three nucleotides called a **codon**. Each codon specifies the incorporation of a given amino acid. Scientists can determine which amino acid a particular codon codes for by finding the codon on a chart called the **Genetic Code** (see Table 7.2).

Overall, the sequence of bases in the DNA is transcribed into the complementary sequence of bases in the mRNA. When the mRNA is threaded through the ribosome, the exposed codons dictate which amino acids will be incorporated into the protein the gene encodes.

Most organisms, from bacteria and fungi to plants and humans, incorporate the same amino acid in response to the same codon. Therefore, two mRNA molecules that carry the same protein-building instructions will be translated to produce the same proteins in two different organisms. Because of this so-called universality of the Genetic Code, bacteria can be used to produce human proteins; plants can produce bacterial proteins, and so on. When an organism is genetically engineered so that it can produce a protein from another species, it is said to be a **genetically modified organism (GMO)**.

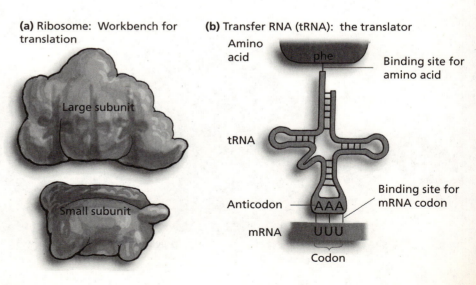

(a) Ribosome: Workbench for translation

Large subunit

Small subunit

(b) Transfer RNA (tRNA): the translator

Amino acid

phe

Binding site for amino acid

tRNA

Anticodon — AAA

Binding site for mRNA codon

mRNA   UUU

Codon

FIGURE 7.2

(c) Translation

1. Amino acids and tRNAs float freely in the cytoplasm.

2. Each tRNA picks up a specific amino acid and carries it to the ribosome.

3. A tRNA will dock if the complementary RNA codon is present on the ribosome.

4. The amino acids link together to form a polypeptide.

Amino acids

tRNA

mRNA

Ribosome

5. The ribosome moves on to the next codon to receive the next tRNA.

6. When the ribosome reaches the stop codon, no tRNA can base pair with the codon on the mRNA and the newly synthesized protein is released.

(d) Termination of translation

7. The chain of amino acids folds into its globular form, and the protein is ready to perform its job.

Protein (such as BGH)

8. The subunits of the ribosome separate but can reassemble and begin translation of another mRNA.

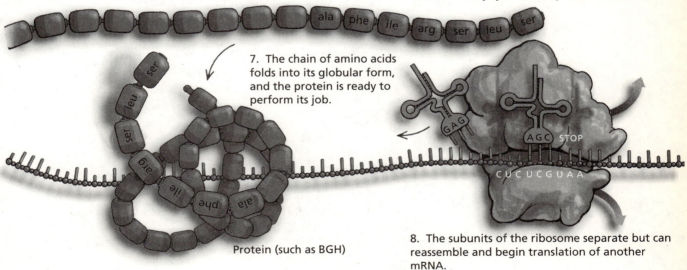

FIGURE 7.2

LAB EXERCISE 7.1

Modeling Transcription

You can think of the information for protein production, which is stored in a DNA molecule, as equivalent to recipes in a cookbook—the recipe is not the food, but the directions for making the food. If you want to eat the dish described, you have to translate words of the recipe into actions.

Transcription of a gene is the first step in converting the information stored in a DNA molecule into a structure that will affect an organism's biology. Using our recipe book analogy, transcribing DNA is a lot like copying a single recipe from a large cookbook onto an index card before you make the dish. Although most cooks probably don't do this, they might if the recipe book was very rare and they wanted to avoid spilling on it, or if the book was enormous and would crowd out all the work space on their counters. The cell transcribes DNA for both of these reasons—the DNA needs to be shielded from damage and it is a huge, unwieldy, molecule.

A recently made transcript is a copy of a single gene whose product the cell currently requires. In human cells, this transcript moves out of the nucleus (where the DNA is stored) and into the cytoplasm where it will be translated.

Look at the double-stranded DNA sequence you have been given. This sequence, which you or one of your classmates built last week, contains the information from a single gene. A sequence of nucleotides on one strand of the DNA double helix serves as a molecular signal that tells the RNA polymerase that this is the beginning of a gene. This sequence of bases serves as a binding site for the RNA polymerase and is called the **promoter**.

Human cells have a promoter at the beginning of each and every gene. The RNA polymerase actually recognizes a sequence of bases in the promoter called the **TATA box**. The term TATA box is an abbreviation for the actual sequence of bases in the promoter to which the RNA polymerase binds, which is TATAAA.

1. Locate the TATA box at the beginning of your gene.
2. Assuming that transcription begins with the first nucleotide after the TATAAA box on the strand complementary to the TATAAA bearing strand, list the sequence of the mRNA that would be produced.

3. The beads representing the RNA nucleotides differ from those representing DNA. Why did you use different beads to represent A, C, and G when making your mRNA transcript?

LAB EXERCISE 7.2

Modeling Translation

The second step in deciphering the code of a DNA sequence is to translate the information from the transcript into a physical structure. The structures coded for by most genes are proteins. Proteins are complex chemicals that comprise the majority of a cell's dry weight. Proteins have many and varied functions. They function inside of cells as enzymes and as structural components of cells. They also help transport other substances into and out of cells.

Regardless of their function, proteins are composed of various amino acids joined to each other. Individual amino acids are joined together to produce a given protein on structures called ribosomes. For this reason, ribosomes are often referred to as the workbenches of the cell. By using the Genetic Code, you can determine the amino acid called for by a given 3-base codon. The Genetic Code is a chart that shows the relationship between the information on the mRNA transcript (or the DNA coding strand it mirrors) and the sequence of amino acids in a protein. The relationship between codon and amino acid is illustrated in Table 7.2.

TABLE 7.2   The Genetic Code

	Second base				
First base	**U**	**C**	**A**	**G**	**Third base**
U	UUU UUC Phenyl-alanine (phe) UUA UUG Leucine (leu)	UCU UCC UCA UCG Serine (ser)	UAU UAC Tyrosine (tyr) UAA Stop codon UAG Stop codon	UGU UGC Cysteine (cys) UGA Stop codon UGG Tryptophan (trp)	U C A G
C	CUU CUC CUA CUG Leucine (leu)	CCU CCC CCA CCG Proline (pro)	CAU CAC Histidine (his) CAA CAG Glutamine (gln)	CGU CGC CGA CGG Arginine (arg)	U C A G
A	AUU AUC AUA Isoleucine (ile) AUG Methionine (met) Start codon	ACU ACC ACA ACG Threonine (thr)	AAU AAC Asparagine (asn) AAA AAG Lysine (lys)	AGU AGC Serine (ser) AGA AGG Arginine (arg)	U C A G
G	GUU GUC GUA GUG Valine (val)	GCU GCC GCA GCG Alanine (ala)	GAU GAC Aspartic acid (asp) GAA GAG Glutamic acid (glu)	GGU GGC GGA GGG Glycine (gly)	U C A G

Note that there is one **start codon** (AUG). This sequence should be present at the beginning of every mRNA. There are also three codons that do not code for amino acids; instead, these codons signal the end of one protein coding sequence and are called **stop codons**.

1. What amino acid should every protein begin with?

(All proteins don't actually begin with this amino acid. It is often cleaved out after protein synthesis.)

2. What sequences make-up the three stop codons?

3. Use the Genetic Code, the blocks representing amino acids, and the string to create a sequence of amino acids that corresponds to your DNA segment. List the amino acid sequence of your protein.

4. Every student at your table received a genetic sequence 39 bases long (after the TATA box).
   A. How many amino acids could the protein specified by this sequence contain?

5. How many does yours contain?

6. Why is there sometimes a difference between the maximum and actual number?

LAB EXERCISE 7.3

Protein Folding and Mutations

After a protein is translated, it folds into a more globular structure than a string because of attraction and repulsion of various amino acids to each other. It is the 3-D shape of the protein that gives the protein its particular function. An analogy of the results of protein folding is the production of a paper boat from a flat sheet of paper. Although its source is a 2-D sheet of paper, by making a 3-D boat, the paper can perform work—that is, it can move along the surface of water and even carry cargo.

The rules for protein folding are still an interesting and active question in biological research. Some rules that are clear are that hydrophobic (water-hating) amino acids minimize their contact with the aqueous cytoplasm of the cell. These amino acids are often found in the interior of a protein. Hydrophilic (water-loving) amino acids will interact with the cytoplasm. They are usually found on the surface of a protein. Oppositely charged amino acids are attracted to each other.

1. Using the following information about the chemistry of some amino acids, transform your string of amino acids into its 3-D shape.

   HYDROPHOBIC: Val, leu, Ile, Met, and Phe
   HYDROPHILIC: Arg, His, Lys, Asp, and Glu-

POSITIVELY CHARGED: Lys, Arg, His
NEGATIVELY CHARGED: Asp, Glu

Changes to the DNA sequence are called **mutations**. When DNA is altered, the mRNA that is produced from the DNA is also altered and changes to protein structure and function can result.

2. Replace one nucleotide on your mRNA and see what effect it has on the protein produced. List the altered nucleotide and amino acid sequences and describe whether the 3-D structure was altered.

3. Is it possible that a change to the DNA would have no effect on protein structure? Why or why not?

4. The addition or subtraction of one or two nucleotides (or multiples thereof) to the DNA results in a **frame shift** mutation. How would this affect the mRNA that is transcribed from the mutated DNA?

5. Frame shift mutations often result in the production of a stop codon where there was not one before. List an altered nucleotide and amino acid sequence that would result in an early stop codon.

LAB EXERCISE 7.4

Labeling GMOs Debate

Genetic modification of food products can involve moving a gene normally found in one organism into another organism. Because of the universality of the Genetic Code, a gene transferred from one organism to another can often be used to build the same protein, regardless of its origin. Foods are genetically modified (GM) to increase their shelf life and to decrease damage from pests and weather.

Concerns about the potential negative environmental and health effects of producing and consuming GM crops have led some citizens to fight for legislation requiring that modified foods be labeled so consumers can make informed decisions about what foods they choose to eat. The manufacturers of GM crops argue that labeling foods is expensive and will be viewed by consumers as a warning, even in the absence of any proven risk. They believe that this will decrease sales and curtail further innovation.

1. Get together with the students at your lab table and discuss the following:
   a. Do you believe that modified foods should be labeled? Summarize your group's discussion.

   b. What you think the potential risks to humans, farm animals, and the environment of GMO consumption and production are. Summarize your group's discussion.

TOPIC 7
_____

POST-LABORATORY QUIZ

TRANSCRIPTION, TRANSLATION, AND GENETICALLY MODIFIED ORGANISMS

1. Using the DNA as a template to make RNA is called _____.

2. The enzyme that uses DNA as a template for the synthesis of a complementary copy of RNA is called _____.

3. Why is the RNA produced by transcription called messenger RNA?

4. Translation occurs on structures called _____.

5. If the DNA from a given gene reads CCATTTGGG, the mRNA transcribed would be _____ and the protein pro-duced would consist of the amino acids _____.

6. How does the nucleotide sequence of the coding strand of a DNA molecule differ from the mRNA produced?

7. The subunits of proteins are _____.

8. What might be the impact of changing the order of amino acids for a given protein?

9. There are hundreds of genes along the length of every human chromosome. What nucleotide sequence would you expect to find at the beginning of each gene?

10. UUU codes for the amino acid phenylalanine in humans and bacteria. This is due to which property of the Genetic Code?

The Theory of Evolution

Learning Objectives

1. Summarize the theory of common descent.
2. Learn how scientists test hypotheses about evolutionary relationships and perform such a test.
3. Define homology and give examples of homologous structures.
4. Use the classification of a number of organisms to create a phylogeny of these organisms.
5. Distinguish between homology and analogy and discuss the role of each in creating and complicating phylogenies.
6. Describe how cladistic analysis is a systematic technique for determining phylogenetic relationships.

Pre-Laboratory Reading

One of the hypotheses Charles Darwin put forth in his book, *On the Origin of Species*, was that all modern organisms derive from a single common ancestor, and that differences between organisms today resulted from evolutionary changes that occurred as species diverged from one another. This once revolutionary idea is referred to as the **theory of common descent**. According to the theory of common descent, all modern organisms can be arranged in a "family tree," or **phylogeny**, that describes their relationship to each other. The theory of common descent is now well accepted as the best explanation for the origins of modern species.

Biologists who study the evolutionary history of living organisms are called **systematists**. Much of what a systematist does is propose hypotheses of relationships among species based on their similarities and differences. One of the generalizations of the science of **systematics** is that organisms that share more features are more closely related than organisms that share fewer features. By this reasoning, systematists think that shared features might have been present in the common ancestor of the organisms. Shared features among species that occur as a result of the species' shared relationships are known as **homologies**. Phylogenies are created by systematists based on their investigation of the homologies among species.

Phylogenies are often presented graphically in the form of a tree diagram (see Figure 8.1). On these **phylogenetic trees**, branch points represent common ancestors to the organisms at the tips of the branches that are connected to them, and branch points near the top of the tree represent more recent common ancestry than branch points near the base of the tree.

Phylogenies derived by systematists represent hypotheses of evolutionary relationship. These hypotheses can be tested using characteristics of the organisms that were *not* used to make the initial hypothesis. This is a bit like coming up with a hypothesis about which foods taste sour and which taste sweet using the appearance of the food, and then testing your hypothesis by measuring the foods' sugar (sweet) and acid (sour) content with some chemical tests. In this case, we can find out if we did a good job of classifying foods by actually tasting them. Systematists cannot, for the most part, know for certain if their hypothesis reflects a true evolutionary relationship, because they cannot travel back in time to observe the actual evolutionary events. However, multiple lines of evidence, including comparisons of DNA sequences and

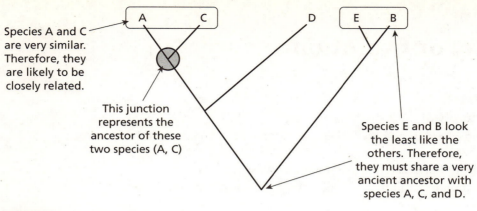

Species A and C are very similar. Therefore, they are likely to be closely related.

This junction represents the ancestor of these two species (A, C)

Species E and B look the least like the others. Therefore, they must share a very ancient ancestor with species A, C, and D.

FIGURE 8.1

investigations of the fossil record, can provide strong support for a phylogenetic hypothesis.

Testing hypotheses of evolutionary relationship is especially important because not all similarities among organisms are homologies. In this case, organisms might appear to share a more recent common ancestor than they actually do. In other words, two organisms that appear similar might not have similar features because they have a common ancestor. Instead, similar environmental conditions experienced by two distantly related organisms might favor the evolution of similar structures or features. Shared characteristics that arise as a result of shared environmental conditions rather than shared ancestry are known as **analogies**. For example, dolphins appear superficially similar to sharks, with their streamlined bodies, lack of body hair, and generally gray color. These similarities are analogous, however, and reflect their similar lifestyles as ocean predators. Dolphins are actually more closely related to other mammals, such as otters and humans, than they are to sharks and fish. Dolphins and other mammals are similar because dolphins breathe air (rather than have gills), produce hair, and provide milk for their infants. Distinguishing between homology and analogy is one of the most difficult tasks a systematist faces.

Cladistic analysis, also known as **cladistics**, is a standardized technique for creating phylogenies. Cladistics allows a systematic comparison between the group of species whose relationship we want to determine (called the **ingroup**) and a species or group of species that is related to this group, but not a part of it (called the **outgroup**). For instance, if we were interested in the possible relationship among frog species, we might choose a toad species as the outgroup. In cladistic analysis, the only useful characters for constructing a phylogeny are those that are found in the ingroup, but not in the outgroup.

LAB EXERCISE 8.1

Examine Homologies Among Skulls or Skeletons

Observe the skulls and/or skeletons of the vertebrate species available. Look for features that some or all of the different specimens have in common. For example, look at the placement of the eyes, the shape of the nose, the type of teeth and their number and distribution in the mouth, and the structure of the bones around the eye socket. For now, we will assume that all similarities are homologies.

Record your observations in Table 8.1, by listing one character (for example, "presence of large canine teeth") in each row and then placing a "+" or a "−" in each corresponding column, indicating the presence or absence of the character.

TABLE 8.1

Character	Species A	Species B	Species C	Species D	Species E	Species F

How many presumed homologies does each animal share with each other animal (Fill in Table 8.2)?

TABLE 8.2

A + B		B + C		C + D		D + E	
A + C		B + D		C + E		D + F	
A + D		B + E		C + F			
A + E		B + F				E + F	
A + F							

LAB EXERCISE 8.2

Generate a Hypothesis of Evolutionary Relationship Among Species Based on Homology

A. Draw a phylogenetic tree.

Remember that one of the generalizations of the science of systematics is that organisms that share more features are more closely related than organisms that share fewer features, and that these shared features were present in the common ancestor of the organisms. Given this generalization, use the data generated in the second table in Lab Exercise 1 to develop a preliminary hypothesis of evolutionary relationship among these animals. Note that it is very likely that you will need to make judgment calls about relationships among some of the species. For example, species A and C might share three homologies, A and B might share three homologies, but B and C might only share two. To resolve the relationship among these three species, you will have to decide which similarities are more likely to be true homologies and which may be analogies.

After you have grouped species into pairs by the number of shared characteristics, group them into larger groups by looking at the number of characteristics each pair shares with another pair. We can assume that these groups of species share a relatively more distant common ancestor.

Draw your proposed phylogenetic tree in the space here. At each branch point on the tree, indicate the homologies that presumably were present in the common ancestor of a group of species. Be prepared to share your analysis with your instructor and classmates.

B. Answer the following questions and be prepared to share your answers with your instructor and classmates.

1. Compare your phylogenetic tree with those created by your classmates. Are there any universal similarities? What is the origin of any key differences?

2. Systematists who specialize in studying the evolutionary relationship among vertebrates use their knowledge of the group to choose characters for phylogenetic analysis and to make assumptions about which shared traits are likely to be homologies and which are likely to be analogies. In general, traits that are highly subject to change via the theory of natural selection (for example, coat color in mammals) are less likely to be homologous among species than traits that are less subject to change (for example, cold-bloodedness). What types of characters in this group might be more useful to creating phylogenies? Why do you think so? What evidence would support the hypothesis that a particular trait contains more information about evolutionary relationship than another trait?

3. The different trees generated by different students in the lab represent alternative hypotheses about the evolutionary relationships among this group of organisms. What additional information would allow you to test these hypotheses and determine which ones are more likely to reflect the true phylogeny?

LAB EXERCISE 8.3

Test a Hypothesis of Evolutionary Relationship by Examining the Classification of the Species in Your Analysis

A. Review the classification system

In the eighteenth century, the Swedish botanist Carolus Linneaus created the modern biological classification system. Within this system, each species is placed within a hierarchy that groups organisms according to ever-broader similarities. The hierarchy takes the following basic form:

Kingdom (broadest classification)

Phylum

Order

Family

Genus

Species (narrowest classification)

Each of these categories is generally called a **taxonomic rank**.

Linneaus' classification system has become the standard method of organizing biological diversity. Since his time, systematists have found it necessary to add "sub" and "super" ranks to help group organisms more finely; for instance, subfamily is a taxonomic rank that falls between family and genus.

Modern classification of species generally uses similarities and differences among modern organisms to determine their places in the hierarchy. Species that share a large number of specific similarities are placed in a narrower taxonomic rank together, while species that share fewer specific similarities, but some broad similarities, are placed in broader taxonomic groups. Note that this is similar to how we grouped organisms using homology in the previous lab exercise. The implication of grouping organisms this way is also similar—organisms in narrower taxonomic groups are presumed to share a more recent common ancestor than organisms in broader taxonomic groups. Therefore, we assume that two species in the same genus share a recent common ancestor, while two groups of genera in the same family share a common ancestor that is slightly less recent.

B. Use the classification of your study organisms to create a hypothesis of evolutionary relationship.

Your lab instructor will distribute the Linnaean classification of each of the organisms you have observed. Use this information and your review of part A of this lab exercise to generate a phylogeny that reflects the evolutionary relationship among these organisms implied by their classification. Draw your phylogenetic tree in this space, and be prepared to share it with your lab instructor and classmates.

C. Answer the following questions:

1. Compare your phylogenetic tree with those drawn by other students in the class. How are they similar? How are they different? What is the origin of these differences?

2. Compare the phylogenetic tree you have created here with the tree you drew in Lab Exercise 2. How are they similar? How are they different?

3. The phylogenetic trees you drew for Lab Exercises 2 and 3, if they are different, now represent alternative hypotheses for the evolutionary relationships among these species. What other information could you use to test which of these hypotheses is more likely to represent the evolutionary relationship? (Note: even if your trees were identical, additional information would help support the hypothesis they represent.)

LAB EXERCISE 8.4

Test Your Hypothesis of Evolutionary Relationship by Comparing DNA Sequences Using Cladistic Analysis

A. Analyze DNA sequence data.

The spreadsheet your instructor has distributed describes a segment of the DNA sequence for a gene that is common to all of the species in this analysis. Although this gene performs a similar function in every animal, slight differences in its DNA sequence among various organisms can provide clues about the possible relationship of these organisms to each other.

On the spreadsheet, a number at the top of each column indicates the position of each nucleotide in the gene sequence. You should notice that all species are identical at many of the positions—for a cladistic analysis, we are only interested in positions where there is a difference in nucleotides among animals. Examine the spreadsheet closely and find the positions where nucleotides are different among members of the ingroup. In Table 8.3, indicate the position number and the nucleotide for each species, including the outgroup, at that position.

TABLE 8.3

DNA Sequence Position	Outgroup	Species A	Species B	Species C	Species D	Species E	Species F

B. Draw a phylogenetic tree based on cladistic analysis of DNA data.

Recall that cladistics analyzes the relationship among organisms in the ingroup by looking at how members of the group differ from the outgroup. In other words, similarities between the outgroup and members of the ingroup cannot provide information about ingroup phylogeny. With this in mind, summarize the information you have collected by indicating the number of **shared differences from the outgroup** for each species pair in Table 8.4.

TABLE 8.4

A + B		B + C		C + D		D + E	
A + C		B + D		C + E		D + F	
A + D		B + E		C + F			
A + E		B + F				E + F	
A + F							

Use the information from this table to construct a new phylogenetic tree, grouping species by the number of shared characteristics. As with your first phylogenetic tree, group pairs of species into larger groups by determining the number of shared differences each pair shares with a different pair. Draw the tree here and be prepared to share it with your lab instructor and classmates.

C. Discuss the following questions and be prepared to share your answers with your lab instructor and classmates:

1. How does this tree compare to your first two trees? Did analysis of DNA data support your initial hypothesis? In what ways did it not?

2. Compare the trees produced by different groups using the DNA data. Do all groups agree on the relationship among these species? Why is this?

3. What other sets of data could you use to test and refine your hypothesis of relationship among these mammals?

LAB EXERCISE 8.5

Wrap-up Discussion

The exercises in this laboratory are directed toward the task of creating and testing a hypothesis of evolutionary relationship among relatively closely related species. However, we can extend these ideas to less obviously similar groups of organisms and think about the data that would allow us to test the hypothesis that organisms as diverse as humans and mushrooms, and indeed all living species, share a common ancestor. Answer the following questions and be prepared to discuss them with your lab instructor and classmates.

1. What basic similarities among mammals and differences between mammals (the ingroup) and reptiles (an outgroup) help support the hypothesis that all mammals share a common ancestor? What additional evidence would help you test this hypothesis?

2. What similarities between mammals and fungi exist? Do these similarities indicate that they both arose from a common ancestor? How could you test your hypothesis?

3. What similarities do all living organisms, from bacteria to dragonflies, share? Are these similarities convincing evidence supporting the theory of evolution? Explain your answer.

TOPIC 8

_____

POST-LABORATORY QUIZ

THE THEORY OF EVOLUTION

1. Describe the theory of common descent.

2. Define "homology."

3. Examine the following phylogenetic tree. Which two organisms probably have the greatest number of homologies? Which two are the descendents, of the most recent common ancestor?

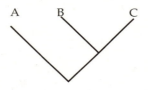

4. Put these taxonomic groups in order from most inclusive to least inclusive.

    Order, Genus, Family, Species, Phylum

5. Draw the relationship among these three organisms that is implied by their classification:
    1. Asparagus. Order Liliales, Family Liliaceae, *Asparagus officinalis*
    2. Yam. Order Liliales, Family Discoreacea, *Discorea species*
    3. Daffodil. Order Liliales, Family Liliaceae, *Narcissus pseudonarcissus*

6. How can a hypothesized phylogeny that is implied by the Linnaean classification of a group of related species be tested?

7. What is the function of the outgroup in a cladistic analysis?

8. Butterflies have wings and bats have wings. Is this similarity a homology or analogy? Explain your answer.

9. Describe two characteristics shared by all living organisms, and which are probably homologous among all organisms.

10. What factors cause some organisms to look superficially similar despite the lack of a close ancestral relationship among them? As part of your answer, give an example of two organisms with analogous traits.

Natural Selection

Learning Objectives

1. Describe the theory of natural selection.
2. Define the terms "fitness" and "adaptation" in the context of the theory of natural selection.
3. Demonstrate how similar populations can become different as a result of natural selection in different environments.
4. Demonstrate how competition for resources might lead to evolution.
5. Use an understanding of the theory of natural selection to explain how a unique feature of a species might have evolved.
6. Relate the origins of variants in a population to the process of mutation.
7. Relate the change in the frequency of an adaptation in a population to a change in the frequency of certain alleles in that population.

Pre-laboratory Reading

Evolution refers to the changes that occur in the characteristics of populations over the course of generations. One of the causes of evolution is the process of **natural selection**. According to the theory of natural selection, the natural variation among individuals within a population of organisms results in differences among them in their ability to survive and/or reproduce in a given environment. Traits that increase the chance of survival and/or reproduction are called **adaptations**. Individuals with adaptations tend to have greater **fitness** than other individuals in the population—meaning that they have more surviving offspring. Many of these offspring carry their parents' adaptations, leading to these adaptations becoming more common in a population and other nonadaptive traits becoming less common. In other words, natural selection results in a change in the characteristics of the population—the population evolves.

The traits that become common in a population through the process of natural selection are unique. This is because adaptations are a function of the environment that the population inhabits. In addition, the traits that become common in a population are limited by the variation that is present in the population. Predators are an environmental factor that causes natural selection—traits that allow individuals to better avoid death through **predation** should spread in a population of prey species. However, many of the traits that are favored by natural selection are subtler in their effect. These are adaptations that improve fitness by making their possessors better **competitors** for a limited resource in the environment. Individuals without such adaptations might not be more likely to die as a result, but they will have fewer successful offspring.

For an adaptation to become common in a population through the process of natural selection, the trait must have a genetic basis. That is, individuals who possess the adaptation must carry a different allele, or set of alleles, compared to those who do not possess the adaptation. The increased fitness of individuals with a particular adaptation means that they produce a large number of surviving offspring who carry the adaptive allele. We can now understand that evolution via the process of natural selection results in a change in the frequency of a particular allele in a population—adaptive alleles become more common and nonadaptive ones become less common.

Understanding the relationship between genes and adaptations also helps us to see how new adaptations arise—through changes in genes, or **mutation**. Mutation is a random process that can have a variety of results—from negative, to neutral, to beneficial. Only a few mutations result in adaptations. In addition, a mutation that results in an adaptation in a particular environment can be neutral, or even harmful, in another environment.

LAB EXERCISE 9.1

Model the Process of Natural Selection Resulting from Predation

A. Simulate the process of natural selection by acting as a "selector," in this case a predator.

The containers of shredded paper available at your lab table represent the habitat (living space) of the prey population. The prey will be represented by pipe cleaners (for prey they are pretty slow, but they *can* hide.)

1. Work in pairs for this exercise. First, decide which member of the pair will act as the "predator" in this habitat and which will be the "prey handler."

2. The predator should turn away while the prey handler hides eight of each color of pipe cleaner within the habitat. Prey do not have to rest on the surface of the paper, but can be buried within the habitat.

3. The predator will now hunt and capture as many prey as possible during a single generation time—in this simulation, 20 seconds. To capture the prey, the predators should use keen eyesight, but also can use their hands to "search" around the habitat—that is, to move paper. When a pipe cleaner is located, it should be placed in the prey container provided.

4. When the time is up, count the number of prey in the container (if a prey was picked up but dropped, it should not be counted—klutzy predators don't eat).

5. Fill in Table 9.1 and add the appropriate number of new prey to the habitat:

   Primary color of habitat _____ _____

TABLE 9.1

	Prey Type 1 _____	Prey Type 2 _____	Prey Type 3 _____
1. Number in habitat at start	8	8	8
2. Number in prey container after hunt			
3. Number remaining in habitat after hunt ($= 1. - 2.$)			
4. New offspring (each survivor has one offspring $= 3.$)			
5. Total number in habitat at beginning of next generation ($= 3. + 4.$)			

6. Perform a hunt of this "second generation" with the same predator and prey handler, record the results in Table 9.2, and add the appropriate number of new prey to the habitat.

TABLE 9.2

	Prey Type 1 _____	Prey Type 2 _____	Prey Type 3 _____
1. Number in habitat at start (= 5. from previous table)			
2. Number in prey container after hunt			
3. Number remaining in habitat after hunt (= 1. − 2.)			
4. New offspring (each survivor has one offspring = 3.)			
5. Total number in habitat at beginning of next generation (= 3. + 4.)			

7. Perform a hunt of the third generation with the same predator and prey handler and record the results in Table 9.3.

8. Remove all prey from the habitat when you are done and post your results on the table provided by your lab instructor.

9. If time allows, switch positions with another pair of students who are using a different habitat type and repeat the three rounds of the simulation (use Tables 9.4, 9.5, 9.6). The predator and prey handler can switch "jobs" for this simulation.

Primary habitat color: _____

TABLE 9.3

	Prey Type 1 _____	Prey Type 2 _____	Prey Type 3 _____
1. Number in habitat at start (= 5. from previous table)			
2. Number in prey container after hunt			
3. Number remaining in habitat after hunt (= 1. − 2.)			
4. New offspring (each survivor has one offspring = 3.)			
5. Total number in habitat at beginning of next generation (= 3. + 4.)			

TABLE 9.4   **Simulation Round 1**

	Prey Type 1 ____	**Prey Type 2** ____	**Prey Type 3** ____
1. Number in habitat at start	8	8	8
2. Number in prey container after hunt			
3. Number remaining in habitat after hunt ($= 1. - 2.$)			
4. New offspring (each survivor has one offspring $= 3.$)			
5. Total number in habitat at beginning of next generation ($= 3. + 4.$)			

TABLE 9.5   **Simulation Round 2**

	Prey Type 1 ____	**Prey Type 2** ____	**Prey Type 3** ____
1. Number in habitat at start ($= 5.$ from previous table)			
2. Number in prey container after hunt			
3. Number remaining in habitat after hunt ($= 1. - 2.$)			
4. New offspring (each survivor has one offspring $= 3.$)			
5. Total number in habitat at beginning of next generation ($= 3. + 4.$)			

TABLE 9.6   **Simulation Round 3**

	Prey Type 1 ____	**Prey Type 2** ____	**Prey Type 3** ____
1. Number in habitat at start ($= 5.$ from previous table)			
2. Number in prey container after hunt			
3. Number remaining in habitat after hunt ($= 1. - 2.$)			
4. New offspring (each survivor has one offspring $= 3.$)			
5. Total number in habitat at beginning of next generation ($= 3. + 4.$)			

10. Remove all prey from the habitat when you are done and post your results on the table provided by your lab instructor.

B. Discuss the following questions and be prepared to share your answers with your lab instructor and classmates:

1. Compare the population of pipe cleaners at the end of the first simulation to the population of pipe cleaners at the beginning of the simulation. Did the population evolve in response to natural selection by the predator?

2. Compare the population of pipe cleaners at the end of first simulation to the population of pipe cleaners at the beginning of simulation. Did the population evolve differently in different habitats? Why?

3. Can you think of prey species in nature that have adaptations that reduce their risk of death by predation? Describe some examples.

4. How would you expect your results (or those of your classmates) to differ if all of the individuals of one of the easy-to-find prey types produced three offspring upon surviving predation? Can you think of prey species in nature that are easy to catch but have high rates of reproduction? Give an example.

LAB EXERCISE 9.2

Model the Effects of Mutation and Competition on Evolution by Natural Selection

A. Simulate competition for food among wading birds.

1. Work in teams of four to complete this exercise. Add to your team's aquarium four of each type of food for each person in your group. Each student should then select an "average length" bird bill (one set of short chopsticks). Practice using these to pick up the "food items" (balloons) *with one hand* until you feel comfortable with how the bill works.

2. Everyone will feed together for a round of 15 seconds, according to the following rules:

   - Food must be picked up by your bill. Do not use the edge of the container to roll the food out of the basin.
   - Pick up only one piece of food at a time.
   - No portion of your hand may enter the water while you are fishing. If it does, you must forfeit the food item.

3. At the end of the round, record the number of pieces of food of each type picked up by each student on Table 9.7.

4. To symbolize the production of the next generation, each student should roll two dice.

TABLE 9.7   **Round 1**

Bird Name	Number of A	Number of B	Total

   If you roll any number besides 7, your offspring will have the same bill type as you. If you roll a 7, your offspring will carry a mutation for bill length. Now roll a *single* die to determine the nature of the mutation, as follows:
   • If you roll a 1 or 2, the mutation results in no change in bill length.
   • If you roll a 3, 4, 5, or 6, the mutation results in a longer bill. Pick up a longer bill and practice with it a few times before the next round.
5.   Do another feeding round following the preceding rules and record the results in Tables 9.8–9.11. Repeat the protocol for determining the traits of the next generation. You should complete five rounds of feeding in total.

TABLE 9.8   **Round 2**

Bird	Beak Length	Number of A	Number of B	Total

TABLE 9.9   **Round 3**

Bird	Beak Length	Number of A	Number of B	Total

TABLE 9.10   **Round 4**

Bird	Beak Length	Number of A	Number of B	Total

TABLE 9.11   **Round 5**

Bird	Beak Length	Number of A	Number of B	Total

B. Answer the following discussion questions and be prepared to share your answers with your instructor and classmates.

1. What is the significance of using a roll of the dice to determine whether mutation occurs? How does this compare to how mutations actually occur?

2. The chance of getting a 7 when rolling two dice is 1 in 6. How do you think this compares to the chance of mutation occurring in any one gene?

3. When a mutation occurs in a "family" (in this simulation, one person represents a "family" of birds), does that change the likelihood that another mutation will occur in that family? Explain.

4. What is the significance of the second roll of the dice (that is, after someone determines that a mutation has occurred)? How does this second roll correspond to the effect of a mutation in nature?

5. Did bill length have an influence on feeding success? How?

6. How might the amount of food an individual consumes affect that individual's reproductive success (that is, number of offspring)? Describe specifically how more food benefits females and how it benefits males.

7. This simulation did not allow for changes in the population due to differences in reproductive success. Given the differences among individuals in bill length and in feeding success, how do you think the population you modeled might have evolved?

8. In the lab introduction, evolution was defined as a change in allele frequency in a population over the course of generations. Did the frequency of the "long bill" allele change in your population of birds (that is, within your laboratory group) over the five generations modeled here? Was this change a result of natural selection? Why not?

9. Does this simulation illustrate how a new species might evolve from an ancestral species? In what way?

10. Comment on how this simulation is a good model of what happens in nature and how it is a poor model of natural processes.

LAB EXERCISE 9.3

Practice Applying Your Understanding of Natural Selection

Each group of students will receive a description of an adaptation in a particular species from the lab instructor. Your task is to discuss this adaptation with your group and describe a reasonable scenario of how it might have evolved in the species. You will present this to the rest of the class. Your presentation should include the following points:

• What were the features of this species before the adaptation arose?
• How did the adaptation likely arise?
• What was the fitness advantage to individuals who possessed this adaptation relative to others who lacked it?
• What are the disadvantages of this adaptation? How might its advantages/disadvantages lead to different outcomes in different environments?

TOPIC 9
_____

POST-LABORATORY QUIZ

NATURAL SELECTION

1. How does natural selection cause evolution?

2. Why don't all populations of a species have the adaptations? (Hint: Think about the exercise where you used pipecleaners as prey animals.)

3. Consider the data in Table 9.12.

TABLE 9.12   Pipe Cleaner Colors

Generation	Green	Brown	White
1	8	8	8
2	10	4	6
3	14	2	4
4	20	2	4

Which individuals have the highest fitness in this population?

4. The pipe cleaners in the simulation of natural selection by predators gave birth to offspring that were the same color. Color in pipe cleaner populations is thus analogous to traits in natural populations that have a _____ basis.

5. Is camouflage the only adaptation for surviving predation? Explain.

6. Traits that increase an individual's success at obtaining limited resources are favored by natural selection in environments where _____ is a major factor.

7. Do all mutations result in adaptations? Why or why not?

8. Chickadees have a remarkable ability to learn from other birds about food sources they have never seen before. In fact, chickadees in the United Kingdom learned how to drink out of milk bottles left on door stoops by milk delivery services. After one or a few chickadees learned how to do this, the ability spread throughout the countryside in a matter of months. Is drinking out of milk bottles an adaptation that is subject to evolution by natural selection in chickadees? Explain.

9. Imagine a dog that is born with a mutation that causes elongated and webbed feet. Is this an adaptation?

10. Use your understanding of natural selection to describe how the following trait might have evolved:

   Plants have pores on the surfaces of their leaves that allow carbon dioxide into the leaf (for photosynthesis), but also allow water to escape. Land plants have guard cells surrounding these pores that actively regulate the size of the pore and will make the pore smaller in dry conditions. The ancestors of land plants did not have guard cells.

TOPIC 10

Species and Races

Learning Objectives

1. Describe the three steps required for the evolution of two new species from a single ancestor.
2. Apply an understanding of speciation to describe the origin of a pair of species.
3. Discuss the types of phenomena that lead to the isolation of gene pools within a species.
4. Restate the definition of biological race provided in the chapter and compare it to the definition of biological species.
5. Test whether populations of humans conform to the biological race concept.
6. Define the theory of convergent evolution and describe why convergent traits are not evidence of biological relationship.
7. Describe the process of sexual selection and compare and contrast this process with natural selection.
8. Describe the phenomenon of assortative mating and discuss how it reinforces human racial differences.
9. Discuss how assuming a biological basis for commonly described human racial categories can support racism.

Pre-laboratory Reading

A **biological species** is a group of individuals that, in nature, can interbreed and produce fertile offspring, but does not breed with members of another species. In other words, biological species are **reproductively isolated** from each other. According to theory, the evolution of two or more descendant species from a single ancestral species—**speciation**—can occur when subgroups, or **populations**, of the ancestral species become isolated from each other. The formation of species generally requires three steps:

1. Isolation of the populations' **gene pools**, that is, the total set of alleles present in the population.
2. The occurrence of evolutionary changes in one or more of the isolated populations.
3. The evolution of reproductive isolation between the populations, preventing the possibility of future **gene flow**, which is the movement of alleles from one population to another.

The gene pools of different populations can become isolated from each other for a number of reasons. A physical barrier might prevent contact between members of the different populations. This barrier might result from a geological event, such as the change in the course of a river that splits a formerly contiguous population, or it might result when a small population of individuals emigrates from the range of the main population. A **temporal**, or time-related, barrier also can stop gene flow among populations of species living in the same area; for instance, if subpopulations of migratory birds arrive at nesting sites at different times in the spring, most of the mating that occurs might be among individuals who have similar arrival times. After their gene pools have become isolated from each other, populations might follow divergent evolutionary paths; for example, if environmental conditions differ for populations in physically isolated environments, **natural selection** might favor one set of

alleles in one environment and a different set in other environments. As isolated populations become more and more divergent from each other, individuals in the different populations might become reproductively incompatible; that is, they might be unable to mate because of large genetic or behavioral differences. Thus, the populations become reproductively isolated and will likely continue to become increasingly different from each other, even to the point of possessing completely different genes.

The period between the separation of gene pools and the evolution of reproductive isolation can be thought of as a period during which **biological races** of a species might form. In other words, races can be thought of as populations of a single species that have diverged, but are not reproductively isolated, from each other. Humans, just like other species, are a product of evolution. Often people assume that this statement implies that differences among groups or "races" of humans are mainly biological in nature; that is, that each race of humans represents a separate, and unique, evolutionary "line" of the human species. However, there is little evidence that "races" of humans are significantly different biologically.

Consider the traits we use to classify humans into racial categories—primarily traits such as skin color, hair texture, and eye shape. Do individuals who share these physical traits also share a common evolutionary history? Perhaps, but we should also consider another factor that causes *unrelated* populations to resemble each other—**convergent evolution**. Convergence occurs when populations are exposed to similar environmental factors and thus experience a similar regime of natural selection. In this case, both populations might *independently* evolve similar solutions to the environmental challenge. This is most easily seen in the similarities between dolphins and sharks. Superficially, dolphins and sharks appear very similar in body shape and color; however, dolphins are mammals and are more closely related to us than they are to their fish look-alikes. Similarities among human populations in skin color also appear to be a result of convergence of populations experiencing similar ultraviolet light levels.

Populations of organisms, especially animals, might also appear different from each other as a result of a process called **sexual selection**. Sexual selection has an effect on traits that influence mating success and is responsible for many of the differences between males and females of the same species. A widespread female preference for a particular trait in a male will cause that trait to become more prevalent in a population—and a widespread male preference is believed to cause females to evolve as well. It appears that some differences among human populations are established by sexual selection and are thus superficial in nature (that is, related to physical appearance only).

Differences among human populations are also reinforced and maintained by **assortative mating**, in which individuals choose mates who are physically and economically similar to them. When most individuals marry within their "racial" groups, the physical differences that help define the races remain distinct.

Natural selection, sexual selection, and assortative mating have led to a variety of differences among human populations—convergent evolution has caused some of these populations to look more similar to each other than to other, equally unrelated, populations. However, as this lab will demonstrate, there are no consistent differences among human populations in different races or consistent similarities among human populations in the same race. "Race" in humans has little or no biological meaning.

LAB EXERCISE 10.1

The Evolution of Reproductive Isolation

Microbotryum is a genus of fungi that causes plant diseases called "smuts." This fungi attacks the male organs (anthers) of affected plants, causing them to blacken and shrivel, resulting in male sterility. Insect pollinators who carry fungal spores from one plant to the other spread this disease. Traditionally, all anther smuts have been assumed to be caused by a single species of *Microbotryum*, but there is a real question about whether anther smut organisms found on different species of plants are in the process of becoming reproductively isolated from each other. This would occur if mating between strains that infect different flowers is unlikely, which might be the case if insect pollinators who carry the fungal spores tend to visit only one species of host flower, or if the host flower populations are geographically separated from each other.

A. Given what you have learned about the theory of how reproductive isolation evolves in separated populations, hypothesize about the degree of reproductive incompatibility among the following strains of *Microbotryum* crossed with the strain that attacks the flowers of *Silene latifolia* in Virginia. Consider both the geographic location of the host and its relatedness to *Silene latifolia* (see Table 10.1). Remember that the theory of evolution states that species in the same genus share a more recent common ancestor (and thus are more closely related to each other) than species in different genera.

TABLE 10.1

Microbotryum Strain Host Plant	Geographic Location of Host	Incompatibility Rank (1 = High, 6 = Low)
Silene latifolia	United Kingdom	
Silene latifolia	Virginia	
Silene caroliniana	North Carolina	
Silene virginica	Virginia	
Paspalum paniculatum	Costa Rica	
Lychnis flos-cuculi	United Kingdom	

B. Test your hypothesis.

1. Work in pairs for this exercise.
2. Obtain a petri dish containing developing fungi. Notice that the dish is marked and divided on the bottom into eight "pie pieces." Seven of these pie pieces contain a different cross between *Microbotryum* strains. The key to the labels is on Table 10.2. (Note: The (A1) and (A2) designations on the table refer to different "sexes" of the fungi. For sexual reproduction to occur, the two strains have to be of different sex.)
3. From each pie section, in turn, use a forceps to lift a small piece of agar that obviously contains fungal growth. Place this fragment face-up on a microscope slide, add a drop of water, and place a cover slip on it. Observe the slide under high power (20X or 40X) on a microscope stage.
4. You should observe many oval sacs on the agar. These are called sporidia. Some sporidia will be conjugating, meaning that a thin tube

will be connecting adjacent sporidia. Ask your laboratory instructor for assistance if you are not certain you have identified a sporidia.

5. Start at one corner of the field of view and count the number of single and conjugating sporidia, up to 100 (or slightly more) total sporidia observed. Count each member of a conjugating pair as 1 sporidia. Enter your data in Table 10.2.

TABLE 10.2

Key	Cross (Name of Host Organism) *Silene latifolia* (A2) x	Number of Single Sporidia	Number of Conjugating Sporidia	Frequency of Conjugation
+C	*Silene latifolia* (A1)			
−C	*Silene latifolia* (A2)			
Sc	*Silene carolinana* (A1)			
SlUK	*Silene latifolia –* UK (A1)			
Sv	*Silene virginica* (A1)			
Sao	*Paspalum paniculatum* (A1)			
Lf	*Lychnis flos-cuculi* (A1)			

6. Calculate the frequency of conjugation by dividing the number of conjugating sporidia by the total number of sporidia observed.

C. Discuss the following questions and be prepared to share your answers with your lab instructor and classmates.

1. Do crosses between different strains show differing amounts of reproductive compatibility?

2. What is the purpose of the "+C" cross (which also could be called the "positive control")?

3. What is the purpose of the "−C" cross (which could be called the "negative control")?

4. Did the results of the experimental crosses support your hypothesis?

5. Explain why the strains of *Microbotryum* found on different plants might be diverging from each other.

6. Do the strains of *Microbotryum* on different plants appear to be different biological species? Do they appear to be different biological races? Support your answer.

LAB EXERCISE 10.2

The Morphological Basis of Human Race Classifications

People often assume that a set of physical (that is, **morphological**) characteristics unite groups of individuals within the major human races. In this series of exercises, you will use a number of photographs of people to investigate whether humans are easily classified into races and whether the morphological differences among these races are as distinct as we might assume.

A. Group people into racial categories.

1. Work together with your lab partners to classify the people in the photographs provided into races.

2. After you have completed your groupings, fill in Table 10.3. Give the groups names (for example, Asians, Hispanics, and so on). For each

TABLE 10.3

Group Name	Physical Traits

grouping, write the name and the *physical characteristics* that the members of the group share, as evidenced by the photographs. Be as specific as possible when describing the physical traits. You might have more or fewer groupings than table rows.

B. Discuss the following questions and be prepared to share your answers with your lab instructor and classmates.

1. Look at the groupings made by other lab tables. Did everyone in the lab group people identically?

2. Discuss the physical traits you used to make your groupings. Were there some key traits that other groups used to classify these individuals that you did not?

3. Were there individuals you found difficult to classify? Why? What additional information would have been useful?

C. Make a closer examination of the traits we use to group people into races.

Take the same pictures you used in part A and arrange them in a line by skin color only. Photos should be arranged to show a gradient from darkest to lightest skin (if people appear to be the same shade, stack the pictures together).

D. Discuss the following questions and be prepared to share your answers with your lab instructor and classmates.

1. Are all the individuals in each race you described in part A next to each other in the line, or are the groups mixed together?

2. Does there appear to be clear "breaks" between people with different skin color types, or is the variation **continuous** (that is, does skin color change gradually as you move along the line)?

3. If you chose another trait to arrange people by (say eye or nose shape), would arrangement of individuals be the same?

4. What does your answer to the previous question tell you about correlation among morphological (physical) characters in people? In other words, do all people with darker skin have one typical nose shape while those with lighter skin have another different shape?

5. The groupings you made in part A of this exercise reflected your feelings about the key morphological traits that define the race of an individual. Why do you think these traits are more important than other traits in determining someone's racial identity?

LAB EXERCISE 10.3

Are Human Races Biological Races?

We can consider the commonly used racial categories (for example, American Indian, Asian, Black, Pacific Islander, White) as a hypothesis; that is, that human populations grouped together in these races are more similar biologically, or share a more recent common ancestor, than human populations in different races. One way to test this hypothesis is to look for nonmorphological similarities among populations within a race. In this exercise, you will investigate whether populations with similar skin colors (and thus assumed to be the same race) are more similar on other, less visible traits.

A. Is skin color a mark of shared evolutionary history?

 1. On the following pages, you will find a series of data tables (Tables 10.4, 10.5, and 10.6) containing information about gene frequencies in different human population groups.

TABLE 10.4   **ABO Blood Types**

People	Place
Low A, no B	
Toba Indians	Argentina
Sioux Indians	South Dakota
Moderate A, no B	
Navaho Indians	New Mexico
Pueblo Indians	New Mexico, Northern Mexico
High A, little B	
Blood Indians	Montana
Australian Aborigines	Southern Australia
Eskimo	Northern Canada
Basques	France and Spain
Shoshone Indians	Wyoming
Polynesians	Hawaii
Fairly high A, some B	
English	England
French	France
Armenians	Turkey
Lapps	Northern Finland
Melanesians	New Guinea
Germans	Germany
High A, high B	
Welsh	Wales, Great Britain
Italians	Italy
Siamese	Thailand
Finns	Southern Finland
Ukrainians	Ukraine
Indians	India

TABLE 10.5 **PTC Tasters**

People	Place
Less than 10%	
Chinese	Taiwan
Cree Indians	Northern Central Canada
Chinese	China
Africans	West Africa
Bantu	Kenya
Lapps	Finland
Japanese	Japan
11–20%	
Chilean	Chile
Malayan	Malaysia
Hindu	Northern India
Cabocio Indians	Brazil
21–30%	
Belgian	Belgium
Portuguese	Portugal
Eskimos	Northern Alaska
Arabs	Sudan
Finns	Finland
Hindu	Southern India
31% and greater	
Norwegians	Norway
English	England
Eskimos	Labrador
Danes	Denmark

TABLE 10.6 **Lactase Deficiency**

People	Place
75% or more are deficient	
Papunya	Australia
Chami	Colombia
Chinese	China
Bantu	Central Africa
Thai	Thailand
Lapps	Northern Finland
Less than 70% are deficient	
Batutsi	Rwanda, Sub Saharan Africa
Eskimos	Greenland
Indian (no tribe designated)	North America
Finns	Southern Finland

2. For each gene described, you should create a different map using the transparency maps available. Color regions of the maps corresponding to the allele frequency group populations in that region. (In other words, all populations in the "High A, High B" group should be the same color, while all populations in the "Low A, No B" group should be a different color.) Overlay these transparent maps and the map of skin color distribution to answer the discussion questions.

B. Discuss the following questions and be prepared to share your answers with your lab instructor and classmates.

1. Does it appear that shared skin color is an indication of other shared nonmorphological traits?

2. Does shared skin color seem to indicate common ancestry? Why or why not?

3. Relate this exercise back to Lab Exercise 2. Does it appear that the groups you made in that exercise reflect distinct biological populations? Why or why not?

LAB EXERCISE 10.4

Discuss Biology and Racism

A. Read the following summary.

In 1994, *The Bell Curve*, a book by Richard J. Herrnstein and Charles Murray, was published. It quickly became a bestseller and the focus of intense controversy. In *The Bell Curve*, Herrnstein and Murray argued that affirmative action programs (that is, programs that preferentially award jobs, scholarships, and so on to members of racial groups that are underrepresented in leadership positions) are doomed to failure because racial groups differ in their innate intelligence. One of the authors' most controversial statements was that black Americans have lower IQs than white Americans, on average, because blacks have genes that result in lower intelligence than whites.

B. Discuss the following questions and be prepared to share your answers with your lab instructor and classmates.

1. Given what you have learned in Lab Exercise 3 about the relationship between shared skin color and other genetic traits, do you think it is reasonable to group all blacks together into one biological group, as Herrnstein and Murray did? Why or why not?

2. Racism is basically the idea that some groups of people are better than others, and that it is somehow justified or proper for the more powerful group to subdue and oppress the less powerful. Is a book like *The Bell Curve* racist? Does it contribute to racist beliefs?

LAB EXERCISE 10.5

Investigate Sexual Selection

As discussed in the pre-laboratory reading, theory states that sexual selection causes the evolution of traits that are related to mating success. Differences among human populations in how individuals choose mates can lead to differences among them in physical traits. However, sexual selection also appears to lead to differences in the preferences and behaviors of males and females in many species.

Sexual selection theory states that males and females should have different reproductive strategies and therefore should favor different characters in their mates. In general, males in species that are polygamous (that is, where individuals have more than one partner) are hypothesized to be primarily interested in the youth and health of their potential mates—younger, healthier females are more likely to produce healthy offspring than older females, and the reproductive success of males is maximized by the total number of offspring they father.

The reproductive success of females is not just a function of the number of offspring they produce, which is limited because producing offspring is so energy intensive for females, but by the "quality" of those offspring in terms of their ability to survive and reproduce themselves. Thus females are hypothesized to be generally more interested in the genetic quality of their mates and the ability of these mates to provide resources for the developing offspring.

Biologists debate whether sexual selection theory can explain the behavior of human males and females, and if it can, how well. In this exercise, you will investigate whether there is evidence that human behavior conforms to sexual selection theory.

A. Do men and women have different preferences for mates?

1. Work in groups of four for this exercise.
2. Given the previous description of how males and females differ, hypothesize about what traits or qualities men would be expected to favor in a potential mate and what traits or qualities women would be expected to favor in a potential mate. List these traits in Table 10.7.
3. Choose two traits from each column in Table 10.7 that you think are likely to be "requested" in personal ads published in standard newspapers. Write these traits in the first column of Table 10.8.

TABLE 10.7

Traits Men Should Favor in Mates	Traits Women Should Favor in Mates

TABLE 10.8

Mate Preferences in Ads	Number of Times Appearing in "Men Seeking Women" Ads	Number of Times Appearing in "Women Seeking Men" Ads
Preferred by Men Seeking Women		
1.		
2.		
Preferred by Women Seeking Men		
3.		
4.		

4. Obtain a page of personal ads from your laboratory instructor.

5. Carefully read the ads in the "men seeking women" and the "women seeking men" categories (or a number specified by your lab instructor, if the listing is very long). If one of the four traits you identified appears in an ad, note this with a check mark in the appropriate cell on the table.

B. Discuss the following questions and be prepared to share the answers with your lab instructor and classmates.

1. Did the results of your survey of the personal ads support the hypothesis that men and women seek different traits in potential mates? Why or why not? Consider the function of personal ads and that the hypothesis applies to polygamous species. Are humans polygamous?

2. Did you notice any additional patterns in the personal ads that might indicate differences between women and men in their mate preferences?

3. Given that differences among human populations are partially a result of differences in traits valued by men and women in each population, do you think personal ads in other cultures might "read" differently?

4. Do you think evaluating personal ads is an adequate way to evaluate the preferences of men and women for mates? Why or why not?

5. If patterns in the personal ads are a reflection of how sexual selection occurs in a modern human population, do you think that men and women are continuing to evolve? Given what you have read, what sorts of traits would you expect to become more common in men? In women?

LAB EXERCISE 10.6

Investigate Assortative Mating

A. Do men and women seek mates who are like themselves?

Personal ads contain information about the physical traits and often the educational/occupational status of the ad's writer as well as the physical traits and educational/occupational status of the mates that are being sought. We can use these advertisements to determine whether men and women desire partners who are similar to them in a number of traits.

1. Work in pairs for this exercise.

2. Obtain a page of personal ads from your laboratory instructor. In many personals, descriptors of race, religion, marital status, and professional status are indicated in a shorthand manner (for example, "BF" might mean "black female"). Find the key that helps explain these categories.

3. Looking at the key, choose three general descriptors (for example, race, divorced versus single, religion specified) that ad writers use to classify themselves and their ideal mates. Write these in the first column of Table 10.9.

TABLE 10.9

General Descriptor	A. Ideal Mate Same as Ad Writer	B. Ideal Mate Different from Ad Writer	C. Did Not Specify Single Preferred Trait in Ideal Mate	D. Total Ads (A + B + C)

4. Examine the ads carefully and record on the table whether advertisers sought mates similar to themselves on the three descriptors you specified previously, different from themselves, or if they did not specify (or specified multiple preferences).

5. Determine the percent of individuals displaying a preference for positive assortative mating for each descriptor by dividing the number in column A by the number in column D. Record this result in Table 10.10.

TABLE 10.10

General Descriptor	Percent with Positive Assortative Mating Preference	Percent with Negative Assortative Mating Preference

6. Determine the percent of individuals displaying a preference for negative assortative mating for each descriptor by dividing the number in column B by the number in column D. Record this result in Table 10.10.

B. Discuss the following questions and be prepared to share your answers with your lab instructor and classmates.

1. Of the three traits you chose for this exercise, which appeared to be most important to advertisers? In other words, which were specified in the largest number of ads?

2. Were advertisers interested in finding mates like themselves or unlike themselves on these important traits?

3. Discuss the traits (if any) that appeared to be less important (as measured by the number of individuals who mentioned the trait). What is the difference between important traits and less important ones? Were individuals showing a positive assortative mating preference for these traits or a negative one?

4. Consider your answers to the previous questions. Does assortative mating appear likely to be reinforcing physical differences between human populations? What about social differences?

TOPIC 10
_____
POST-LABORATORY QUIZ

SPECIES AND RACES

1. In the cross of fungal strains of different sexes found on different hosts, what observation would convince you that these two strains are actually different species?

2. Given the results of the crossing experiment between different strains of anther smut, how might you expect the outcome of crosses between individuals of different human races should turn out, if these races represent true biological races?

3. Skin color in humans is continuously variable. What does this mean?

4. Is shared skin color among human populations correlated to other shared morphological traits? Explain your answer.

5. Are populations of humans who share the same skin color similar "underneath the skin" as well? Explain your answer.

6. Human populations that share a similar skin color might appear similar as a result of convergent evolution. What is convergent evolution?

7. Why are females expected to prefer different traits in their potential mates than males prefer in their mates?

8. If modern humans choose their mates as predicted by sexual selection theory, what sorts of traits should women desire in men?

9. Does positive assortative mating for skin color and ethnic background reinforce or dilute differences among human races? Explain your answer.

10. What is racism?

TOPIC 11

Sexually Transmitted Diseases

Learning Objectives

1. View several different shapes of bacteria and some examples of bacteria that cause sexually transmitted diseases.
2. Measure the effectiveness of antibacterial products.
3. See how disease can spread throughout a population.

Pre-Laboratory Reading

Infectious diseases occur when an infectious agent gains access to the host organism's body and uses host resources for its own purposes. Infectious diseases are transmitted via contact with the disease-causing organism. Disease causing organisms include bacteria, viruses, protozoans, fungi, and insects. This laboratory will focus on sexually transmitted diseases caused by bacteria, protozoans, fungi, and insects because viruses are too small to view with the microscopes that are generally available in teaching laboratories.

Bacteria are single-celled prokaryotes. **Prokaryotes** have no nucleus or membrane-bound organelles. **Protozoans** are also single cells (usually), but they are **eukaryotes** and therefore have a nucleus and membrane-bound organelles. **Fungi** can be single cells or multicellular eukaryotes that resemble plants, but don't have chlorophyll. **Insects** are small animals without backbones (invertebrates). When these organisms cause disease, they are called **pathogens**.

Not all bacteria are pathogens. In fact, your body is the home to many beneficial bacteria. Pathogenic bacteria can cause several sexually transmitted diseases such as Chlamydia, Gonorrhea, and Pelvic Inflammatory Disease.

Chlamydia is caused by the bacterium *Chlamydia trachomatis*. Symptoms of Chlamydia can include pelvic pain and fluid discharge. Unfortunately, many people with Chlamydia experience no symptoms at all and the infection goes untreated. Untreated Chlamydia infection can lead to Pelvic Inflammatory Disease and can result in infertility. Abstinence and condoms will prevent transmission of this disease.

Gonorrhea is caused by the bacterium *Neisseria Gonorrhoeae*. Symptoms can include a thick discharge from the penis or vagina. However, many people experience no symptoms and this infection often goes untreated. Untreated Gonorrhea can cause infertility in women if bacteria spread to the oviducts and cause Pelvic Inflammatory Disease. Abstinence and condoms prevent transmission.

Pelvic Inflammatory Disease (PID) develops when an infection spreads to the uterus and oviducts. Symptoms, when present, include pelvic pain and difficulty becoming pregnant due to scarring and blockage of reproductive organs caused by the infection. Abstinence and condoms prevent transmission.

Bacterial infections can be treated with antibiotics but these drugs might not kill all the bacteria present. Any bacterial cell with a preexisting resistance to the antibiotic will survive and reproduce rapidly because there is less competition from other bacteria. The resistant bacterial cells will pass on their resistance to their progeny, resulting in a large population of resistant bacteria.

Bacterial growth is exponential; that is, one cell gives rise to two, and those two yield four, the four divide to become eight cells, and so on. Exponential division results in rapid bacterial growth.

Many sexually transmitted diseases are transmitted by contact with fluids that carry the pathogen. Transmission can occur via contact with saliva, vaginal fluids, and semen. Other sexually transmitted diseases are transmitted by direct contact with the infectious organism, as is the case with Trichomoniasis, pubic lice, and yeast infections.

Trichomoniasis is caused by the parasitic protozoan *Trichomoniasis vaginalis*. This disease is transmitted via sexual intercourse. The major symptom of Trichomoniasis infection in women is vaginal itching with a frothy yellow-green vaginal discharge. Most men do not have symptoms, but some might experience irritation in their urethra after urination or ejaculation. Abstinence and condoms prevent transmission.

Pubic lice are transmitted through skin-to-skin contact or contact with an infected bed, towel, or clothing. The most common symptom of pubic lice infection, also called crabs, is itching of the pubic area. The itching is caused by an allergic reaction to the bite. This symptom starts about five days after the initial infection. Pubic lice are killed by washing the affected area with a delousing agent. Condoms will not prevent transmission of pubic lice.

Yeast infections are caused by fungi of the genus *Candida*. These yeast are normal inhabitants of the female reproductive tract. They increase in number when a woman is weakened by illness or upset by stress. Antibiotics, taken to treat bacterial infections, can kill vaginal bacteria and allow the yeast to grow, leading to yeast overgrowth. Yeast also can be passed from person to person, such as through sexual intercourse. Yeast infections are characterized by a thick whitish discharge from the vagina and vaginal itching. Men can carry the yeast in their urethra and pass the yeast on during sexual intercourse. Abstinence and condoms help prevent transmission.

An **epidemiologist** is a scientist who attempts to determine who is prone to a particular disease, where risk of the disease is highest, and when the disease is most likely to occur. Epidemiologists try to answer these questions by first determining the source of the infection.

LAB EXERCISE 11.1

Viewing Pathogens

A. Identify bacterial shapes.

Bacteria can be rod-shaped (bacilli), spherical (cocci), or spiral-shaped (spirochetes).

1. View the prepared slides of bacteria labeled A, B, and C under the microscopes. Which sample preparation shows cocci, spirals, and rods?

A = _____
B = _____
C = _____

B. Identify bacteria that cause sexually transmitted diseases.

1. View the prepared slides of the bacteria that cause Gonorrhea and Chlamydia. Sketch these bacterial cells here.

2. Are these bacteria bacilli, cocci, or spirochetes?

C. Identify insect, protozoan, and fungal sexually transmitted diseases.

1. View the prepared slides of the insect that causes pubic lice (*Pediculus pubis*), the protozoan that causes Trichomoniasis (*Trichomoniasis vaginalis*), and the fungus that causes yeast infections (*Candida albicans*).

2. Sketch these organisms here.

LAB EXERCISE 11.2

Testing the Effectiveness of Antibacterial Products

A. Measure the effectiveness of antibacterial agents.

1. Look at the plate of bacteria containing circular pieces of filter paper soaked in various antibacterial agents. The clear area around the disk is devoid of bacteria. The absence of bacteria is due to antibacterial agent present on the disk. Measure the clear area around each disk.

2. Which of the agents tested was the most effective at killing bacteria?

3. Is killing bacteria always a good idea? Why or why not?

4. If the plate had been seeded with viruses instead of bacteria, would you expect the same results as you obtained with the plates of bacteria? Why or why not?

LAB EXERCISE 11.3

Disease Transmission

A. Determine how disease spreads in a population.

This exercise will demonstrate how disease can quickly be spread throughout a population and how epidemiologists try to track the source of a disease.

1. Locate the numbered test tubes on your lab table. Assign one test tube to each person at your lab table. The fluid in the test tube represents your bodily fluids. The bodily fluid in question depends on the disease that we're modeling. In the case of a cold, the fluid in the test tube would represent mucus and saliva. In the case of AIDS, the fluid in the test tube would represent blood and semen. Write the number from the test tube you were assigned on the following data sheet.

2. You will be using your test tube to exchange fluids with four other students. Each time you meet another person, you can exchange bodily fluids by giving them a dropper full of your fluids and taking a dropper of theirs. Gently swirl your test tube after every exchange. Record your swapping partner's tube number on the data chart. It is important

that you keep track of the order in which you made your exchanges. Make sure that you mingle with all of your classmates, not just the other students at your table.

3. When you have finished exchanging, bring your test tube back to your lab table. You will now test your fluids for disease. Add 4–5 drops of phenol red to your test tube. If the contents of the tube turn pink, you have escaped infection. If they turn yellow, you have been infected. Record your infection status on the data sheet. Empty out the contents of the small test tube and rinse both the test tube and eye dropper well.

   Your Source Tube Number _____

Exchange #	Name of Partner	Partner's Source Tube Number
1		
2		
3		
4		

   Were you infected? _____

4. Work with your lab table group to determine the source of infection in Part A. It is easiest to use the information in the format that is posted on the blackboard. You will at best be able to narrow down the possible sources to two, even though only one individual was infected to begin with. Who are the likeliest sources of your infection?

5. Only one person had the infection at the beginning of this exercise. After one exchange, two people were infected (the original carrier, and the person the carrier exchanged with). What is the total possible number of people infected by the end of four exchanges (that is, if an infected person always exchanges with a noninfected person)?

6. What term could be used to describe the rate at which the infection spread?

7. Based on this understanding of infection spread, do you think it is more effective to concentrate money and effort on cures/treatments for infectious diseases or on slowing down or preventing the spread of an emerging disease? Why?

TOPIC 11
_____

POST-LABORATORY QUIZ

SEXUALLY TRANSMITTED DISEASES

1. What kinds of organisms are often human pathogens?

2. Why did we look at prepared slides of pathogens instead of studying live organisms?

3. What types of bacteria cause Pelvic Inflammatory Disease?

4. Infection with Gonorrhea and Chlamydia often goes untreated. If a woman has an untreated infection that advances to PID, why can't antibiotic treatment necessarily prevent her from becoming infertile?

5. What danger is inherent in the use of antibacterial products?

6. How do yeast infections arise?

7. Can men transmit yeast infections to women?

8. Why does having pubic lice make the pubic area itchy?

9. If a person did not exchange fluids from their test tube during Lab Exercise 3, is there any chance they could become infected?

10. Is it always possible to determine who the first person to transmit an infectious disease was?

Development and Sex Differences

Learning Objectives

1. Understand fertilization and witness fertilization in a model organism.
2. Determine the biological mechanism of action for various birth control methods.
3. Measure skeletal differences between human females and males.
4. Test a hypothesis about sex differences in athleticism.

Pre-laboratory Reading

Human males produce sperm cells in their testes and females produce egg cells in their ovaries. Sperm and egg are both examples of **gametes**. Gametes carry one half the number of chromosomes that other body cells carry. This decrease in chromosome number is accomplished when the process of meiosis occurs in the testes and ovaries.

Women undergo meiosis to produce egg cells. Typically, one egg cell per month leaves the ovary and is drawn into the oviduct. Sperm produced in the testes of males must travel out of the penis and through the vagina, cervix, and uterus of a female to reach the oviduct. If an egg cell has been ovulated and sperm are present in the oviduct, fertilization can occur. After fertilization, the egg cell rolls down the oviduct and implants in the lining of the uterus, where it begins to make copies of itself by undergoing the process of mitosis. The fertilized egg cell eventually gives rise to a multicellular organism.

Fertilization can be prevented by abstinence or by the use of different birth control methods. Some birth control methods, called barrier methods, prevent pregnancy by preventing sperm and egg contact. Other birth control methods kill sperm, thicken the woman's cervical mucus, or prevent ovulation. Surgery also can be performed to disrupt the ducts that sperm and egg travel through, permanently preventing fertilization.

When fertilization does occur, different developmental pathways are followed by human females and males. Embryonic developmental differences result in the production of different gonads (ovaries in females and testes in males), different gamete carrying structures (oviducts in females and the epididymis in males), and different external genitalia (the vulva in females and the penis in males).

At puberty, the gonads function as **endocrine organs** by secreting specific hormones. Estrogen secreted in females leads to increased fat storage. Testosterone secreted by males leads to increased muscle mass. Some skeletal structures are larger in males to accommodate their larger muscles. Males also tend to be taller and have longer legs and arms than females because they begin puberty later than females. Because appendages grow at a rate that is proportional to the torso length, a later onset of puberty means the male torso will be longer when puberty begins, resulting in greater limb length. These differences lead to differences in female and male centers of gravity, with the female center of gravity being lower in the body.

The pelvic bones also differ in females and males. Females have more broad pelvic bones (ossa coxae) and a larger rounder pelvic inlet. The female pelvis is also flatter, more broad and tipped more forward than the male pelvis, leading to a sharper angle between the hip and knee (the Q angle).

LAB EXERCISE 12.1

Sea Urchin Fertilization

A. Fertilize sea urchins.

1. View the sea urchin egg under the microscope.
2. Add a drop of sperm to the egg while watching the egg cell under the microscope under low light. Fertilization occurs quickly. You can see that fertilization has occurred when there is a halo around the egg. This halo is actually the fertilization membrane.
3. Turn off the microscope light so that you don't dry out the slide and check back every 10–15 minutes to see if the fertilized egg has begun to undergo cell division. This usually takes from 45–60 minutes for the first division to occur. If your egg cell begins to dry out, add a drop of seawater.

   While you wait for your egg cell to divide, perform Lab Exercises 2–4.

LAB EXERCISE 12.2

Birth Control Methods

A. Study birth control methods.

1. On the lab bench are various examples of birth control devices and pharmaceuticals.

   List the biological mechanism by which each of these methods prevents pregnancy.

   a. serves as a barrier to sperm and egg contact
   b. kills sperm
   c. progesterone causes thickened cervical mucus which prevents fertilization
   d. continuous doses of estrogen and progesterone prevent ovulation
   e. prevents uterus from supporting pregnancy

2. Arrange these birth control devices from least effective to most effective. Failure rates represent the percentages of users per year who have an unintended pregnancy.

   Check with your laboratory instructor or Table 12.1 in your text book to determine whether your predictions are correct.

LAB EXERCISE 12.3

Sex Differences in Skeletal Structure

Average sex differences in skeletal structure exist in humans. These sex differences include several differences in the skull, pelvis, Q angle, and center of gravity.

A. Determine sex differences in the human skull.

1. Compare the skull models from a human female and human male located on the display table. These skulls are labeled A and B. Which skull do you think is a model of a human male and why?

B. Determine sex differences in the human pelvis.

1. Find the two human pelvis models located on the display table. These pelves are labeled A and B. Which pelvis do you think is a model of a female pelvis and why?

C. Determine sex differences in Q angle.

   The Q angle is the angle formed between the hip bone, knee cap, and foot. To measure your Q angle, follow these steps:

1. Stand up.
2. Place the rounded portion of the goniometer on your knee cap.
3. Point the upper portion of the goniometer through your femur toward your hip bone.
4. The lower portion of the goniometer should point along the line made by your tibia (shin bone).

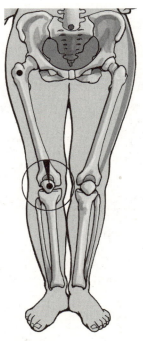

FIGURE 12.1

5. What is your Q angle?

The average male has a Q angle of around 13 degrees; females are around 18 degrees.

6. Why might males, on average, have smaller Q angles than females?

D. Determine sex differences in center of gravity

1. Determine the weight of the board by adding together the readings on both scales. Record this weight in Table 12.2.

TABLE 12.2

Weight of Board (Right Scale + Left Scale)	Length of Board in Inches	Right Scale	Left Scale	Your Height in Inches

2. Determine the length of the board by measuring the distance between the two rulers. Record the length in the table.
3. Lie on the board, with your feet slightly above the ruler on the right scale, and your head slightly above the ruler on the left scale. Hold your arms close to your body.
4. Have your lab partner or the lab instructor read both scales and record the weights on each scale in the table.
5. Follow steps 1–4 to calculate your center of gravity.

    1. Subtract half the weight of the board from each recorded value to find the new value.
       Right scale – $\frac{1}{2}$ the weight of the board = _____ adjusted weight right
       Left scale – $\frac{1}{2}$ the weight of the board = _____ adjusted weight left
    2. $\underline{\hspace{3cm}} \times \underline{\hspace{3cm}} = \underline{\hspace{3cm}}$
       Length between scales   adjusted weight left
    3. $\underline{\hspace{3cm}} + \underline{\hspace{3cm}} = \underline{\hspace{3cm}}$
       adjusted weight left   adjusted weight right
    4. $\underline{\hspace{3cm}} \div \underline{\hspace{3cm}} = \underline{\hspace{3cm}}$
       Product of step 2   sum of step 3   Center of gravity

6. To find your body's center of gravity, hold one end of the tape measure on your toes and measure up the number of inches calculated in step 4.
7. Why might a male and female of similar height have a different center of gravity?

LAB EXERCISE 12.4

Sex Differences in Athleticism

Sex differences in athleticism can be biological or cultural in origin. During this portion of the laboratory exercise, you will be measuring each student's ability to throw a yard dart accurately to determine if sex differences exist for this skill.

1. Devise a hypothesis about targeting ability in females and males.

2. Test your hypothesis as follows:
    This task involves the overhand throwing of darts into the circle. Standing with your toes on the tape line, each member of your lab group should throw three darts with his or her right hand and then three darts with his or her left hand. Award each thrower 2 points for landing in the dart circle and one point for landing outside the dart circle, but inside the tape square. Remove the dart after each throw so that landed darts do not obstruct other darts. Points are awarded for initial landing only. Darts carried into the square or circle by momentum are awarded 0 points. List your group member's scores here:

3. Write the scores for each group member, according to their sex, on the blackboard.

4. After the entire laboratory section has listed its scores on the blackboard, calculate the **mean** or average for each sex by adding all the scores together and dividing by the number of participants in each group.
    Mean for females = _____
    Mean for males = _____

5. Using the data from the entire section, determine the range (lowest to highest score) of scores for each sex and whether the ranges overlap.
    Range for females = _____
    Range for males = _____
    Overlap of ranges = _____

6. Does the data gathered by the class support your hypothesis? Why or why not?

7. Would it be possible to predict the sex of an individual based on his or her targeting score? Why or why not?

8. Are there sex differences in targeting performance between the females and males of this laboratory section? If so, why do you think these differences exist?

9. Skill at targeting would be developed by practicing many different sports. Find out if your lab mates have played sports that would help them develop this skill. Is there a difference in the amount of time one sex spent practicing this skill?

TOPIC 12
_____

POST-LABORATORY QUIZ

DEVELOPMENT AND SEX DIFFERENCES

1. Describe the process of fertilization in humans.

2. List several barrier methods of birth control.

3. Which birth control methods also offer protection from sexually transmitted diseases?

4. How does the birth control pill prevent pregnancy?

5. How do birth control methods that supplement progesterone prevent pregnancy?

6. Why does the male skull have some larger bones?

7. What causes females to have a larger Q angle than males?

8. What factors need to be taken into account when determining center of gravity?

9. Why is the female pelvic inlet more rounded?

10. How do overlapping ranges impact one's ability to make predictions about groups of people?

TOPIC 13

The Human Nervous System

Learning Objectives

1. Describe the components of the nervous system.
2. Describe the transmission of a nerve impulse.
3. Describe how psychoactive drugs affect the nervous system.
4. Understand the functions of the general senses.
5. Delineate the difference between a reflex and an integrated response to a stimulus.
6. Apply the scientific method to questions about senses and learning.

Pre-laboratory Reading

Every second, millions of signals are generated by stimuli that help indicate your body's movements and environment. Your **nervous system** (see Figure 13.1) interprets these messages and decides how to respond. Interpreting and responding to stimuli requires the action of specialized cells called **neurons** (see Figure 13.2). These cells are often bundled together, producing structures called **nerves**. The nervous system consists of the brain, the spinal cord, the sense organs, and the nerves that link these organs together.

Information about your body's movements and environment—sensory input—is detected by **sensory receptors**. When the neurons of these receptors are stimulated, signals are generated and carried to the brain for decoding and integration with other information. Our senses consist of five **general senses** and five **special senses**. The general senses, receptors for which are scattered around the body, are temperature, touch, pain, pressure, and body position. The special senses, receptors for which are found in specialized **sense organs**, are smell, taste, hearing, vision, and equilibrium.

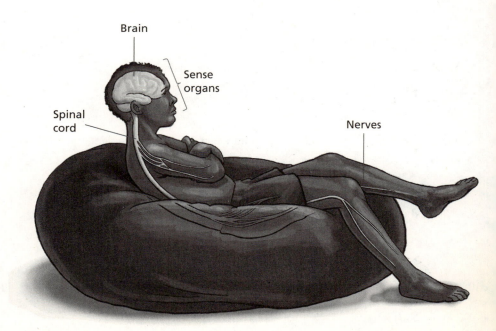

FIGURE 13.1

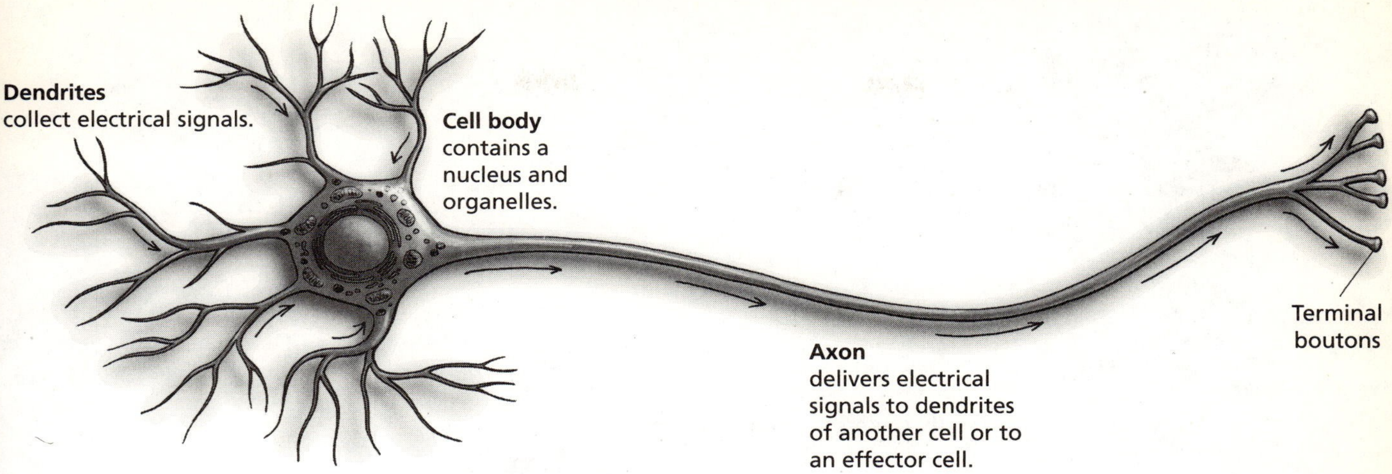

Dendrites
collect electrical signals.

Cell body
contains a
nucleus and
organelles.

Axon
delivers electrical
signals to dendrites
of another cell or to
an effector cell.

Terminal
boutons

FIGURE 13.2

Although the brain is the primary organ of sensory integration, some responses to stimuli do not require its involvement. These responses are called reflexes and are "prewired" in a circuit of neurons called a reflex arc, consisting of a **sensory neuron** that receives information from a sensory receptor, an **interneuron** that passes the information along, and a **motor neuron** that causes a muscle response. Reflexes cause you to react quickly to dangerous stimuli; for instance, the withdrawal reflex occurs when you touch something hot.

To carry information between parts of the body, the cells of the nervous system pass electrical and chemical signals to each other. Information is carried along nerves by electrical changes called **nerve impulses**. Nerve impulses are transmitted between neurons and from neurons to other cells by chemical stimuli, called **neurotransmitters**, released from the neurons.

A nerve impulse results from a small electrical change that is conducted along the length of a neuron. The inside of a neuron is negatively charged relative to the outside as a result of a difference in the concentration of positively charged sodium ions inside and outside of the cell. This difference is maintained by the action of enzymes in the cell membrane of a neuron, which actively pump sodium from the cell. These ions can re-enter the cell only through specialized pores called **sodium channels**. When a neuron is stimulated, sodium channels open and positively charged ions flow into the cell. A domino effect known as an **action potential** occurs as the positively charged sodium ions move toward neighboring sodium channels, causing them to open and let in additional sodium. The sodium ions also attract other, negatively charged ions from elsewhere in the cell; this effectively neutralizes the positive charge at the original site of the stimulus, causing the sodium channels to close. In this way, a nerve impulse is transmitted in one direction along the length of the axon of a neuron (see Figure 13.3).

Adjacent neurons are not directly attached to each other; they are separated by small gaps called **synapses**. When a nerve impulse reaches the end of a neuron, neurotransmitters are released that chemically conduct the impulse across the synapse. Neurotransmitters released by one cell bind to specialized cell receptors on the membrane of an adjacent cell (see Figure 13.4). This binding stimulates a change in the uptake of sodium, and thus triggers an action potential in the receptive cell.

After the neurotransmitter evokes a response, it is removed from the synapse. Enzymes degrade some neurotransmitters, while others are reabsorbed

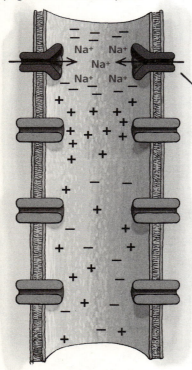

Nodes
of
Ranvier

Nerve cell

(b) Generation of nerve impulse

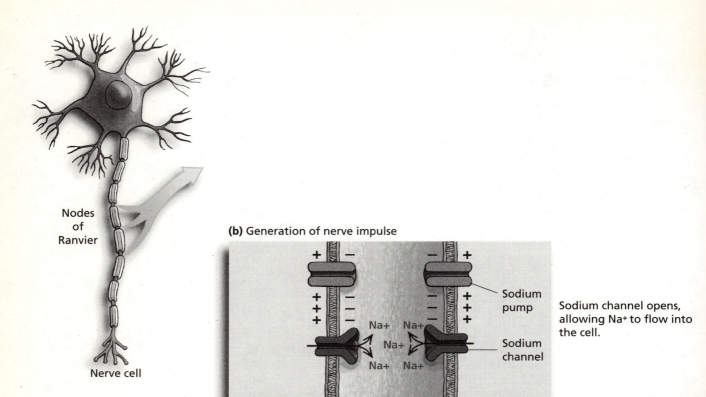

Sodium
pump

Sodium channel opens,
allowing Na+ to flow into
the cell.

Sodium
channel

(c) Propagation of nerve impulse

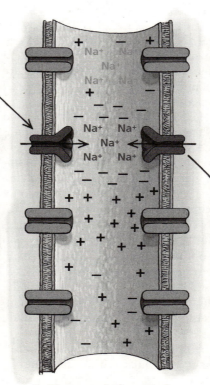

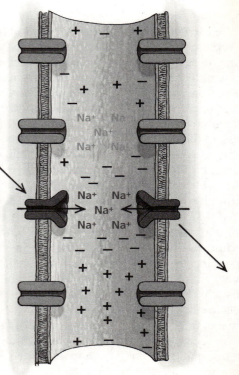

1. Nerve impulse starts with inflow of Na+ ions, which attract negatively charged ions and repulse positively charged ions.

2. The spread of positively charged ions toward the next sodium channel causes it to open, allowing the Na+ to rush in.

3. The depolarization passes down the axon, propagating the nerve impulse.

FIGURE 13.3

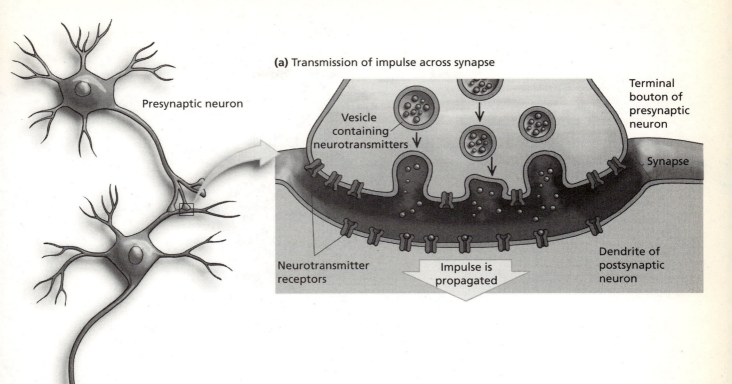

(a) Transmission of impulse across synapse

Presynaptic neuron

Terminal bouton of presynaptic neuron

Vesicle containing neurotransmitters

Synapse

Neurotransmitter receptors

Impulse is propagated

Dendrite of postsynaptic neuron

Postsynaptic neuron

FIGURE 13.4

by the neuron that secreted them, a process called **reuptake**. The rapid removal of neurotransmitters from a synapse prevents continued stimulation of a nerve.

Neurotransmitters play an important role in coordinating the responses of the brain to outside stimuli. The brain is organized into three major regions: the cerebrum, the cerebellum, and the brain stem, each with a different role in processing and controlling bodily functions (see Figure 13.5). Many **psychoactive** drugs are either mimics of brain-specific neurotransmitters or affect the release, degradation, or reuptake of these neurotransmitters. By affecting the activities of neurons in the brain, psychoactive drugs have the ability to change an individual's feelings, perception, and ability to integrate and respond to stimuli. For example, cocaine decreases the reuptake of the neurotransmitters norepinephrine and dopamine, leading to a rush of intense pleasure (a result of increased dopamine activity), and increased physical vigor (a result of increased norepinephrine activity).

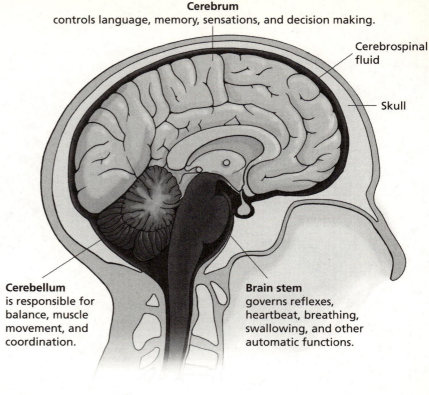

Cerebrum
controls language, memory, sensations, and decision making.

Cerebrospinal fluid

Skull

Cerebellum
is responsible for balance, muscle movement, and coordination.

Brain stem
governs reflexes, heartbeat, breathing, swallowing, and other automatic functions.

FIGURE 13.5

LAB EXERCISE 13.1

Model Transmission of Nerve Impulses Within and Between Neurons

The axon of a neuron is "insulated" with a material called the myelin sheath, made up of Schwann cells (see Figure 13.6). When a nerve impulse travels along an axon, the electrical signal jumps the gap between adjacent cells in the sheath. This process is called saltatory conduction.

(b) Nerve impulses travel more quickly on myelinated nerves.

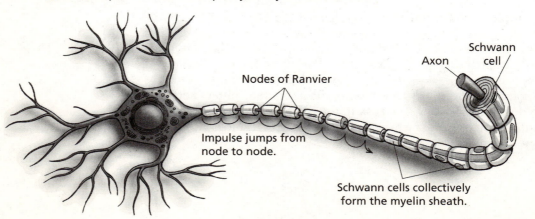

Nodes of Ranvier

Axon

Schwann cell

Impulse jumps from node to node.

Schwann cells collectively form the myelin sheath.

FIGURE 13.6

A. Work with five other students to role play the process of saltatory conduction along a single neuron. The end result of the transmission of an impulse from one end of the neuron to the other is the release of a neurotransmitter, symbolized by the balls. Use your imagination and creativity! Consider how you might demonstrate the propagation of a nerve impulse on the unmyelinated sections of an axon (review Figure 13.3), and how you might symbolize the intervening myelinated sections. Share your role play with your lab instructor and the other students in the class.

B. Determine the length of time required for a nerve impulse to travel along your "neuron."

1. Measure the total length of the axon as you and your teammates are modeling it.

    = _____meters

2. Time from initial nerve impulse until release of neurotransmitter

    = _____seconds

3. Determine the rate of transmission by dividing the length by the time.

    = _____meters/second

C. Now you and your teammates should role play the transmission of nerve impulses among neurons via the use of neurotransmitters. Remember that neurotransmitters do not remain in the synapse for long and are either degraded or taken up again by the neuron that released them. In this role play, each individual is a single neuron—the balls still symbolize neurotransmitters. Share your role play with your lab instructors and other students.

D. Discuss the following questions and be prepared to share your answers with your lab instructor and classmates:

1. How did the fastest rate of transmission your teammates generated in the first role play compare to the fastest neurons, which transmit at 200 meters/second?

2. Not all neurons respond to the same neurotransmitters. How could you have modeled this in your second role play?

LAB EXERCISE 13.2

Discuss Human Modifications of Neurotransmission

A. Discuss the following questions at your lab table and be prepared to share your answers with your lab instructor and other students in the class:

1. Some psychoactive drugs are illegal, such as cocaine and heroin; others are legal, but only obtained via a doctor's prescription, such as Ritalin and Prozac, and still others are legal and available over the counter, such as nicotine. Do you think there should be a set standard for characteristics of psychoactive drugs that can be legally obtained and those that should be illegal (for example, degree of brain function modification, side effects, risk of addiction)?

2. Many experts on drug addiction call this condition a "brain disease." Repeated drug use actually changes the brain—changes ranging from fundamental and long-lasting changes in the biochemical makeup of the brain, to mood changes, to changes in memory processes and motor skills. These changes have a tremendous impact on all aspects of a person's behavior. For instance, drugs that stimulate neurons in the brain's "pleasure centers" actually dull these neurons to normal levels of "pleasure" neurotransmitters and make the user less able to appreciate normal positive life events. Does the fact that drug use changes the structure of the brain, leading to addiction, affect your feelings about the safety of the recreational use of addictive drugs? Does it affect your feelings about those who are addicted? Explain your answer.

3. Attention Deficit Disorder (ADD) and chronic depression are also often considered brain diseases. These two conditions are commonly treated with psychoactive drugs. Some critics of drug therapy for these conditions argue that we are drugging people for "being human" and that our society's acceptance of depression and ADD as "illnesses" is in stark contrast to the general feeling that addiction to illegal drugs is a problem of "personal responsibility," when the science shows that they are essentially two sides of the same coin. Do you agree with these critics? How else might we, as a society, deal with ADD and depression?

LAB EXERCISE 13.3

Brain Structure and Function

A. Work in pairs for this exercise.
B. Spray the brain mold with a light coating of vegetable oil.
C. Mix the following ingredients together in the 1000ml beaker, and pour into the brain mold:

1. 240 ml clean sand
2. 180 ml instant potato flakes
3. 300 ml *hot* tap water

D. Let the mixture set for a few minutes, and then carefully flip the mold onto the tray provided to remove the "brain." You now have produced a model of one hemisphere of the human brain that is fairly accurate in size, weight, consistency, and structure!

E. Identify the three major brain regions on the model: cerebellum, cerebrum, and brain stem. Refer to Figure 13.5 or models available in the laboratory for guidance.

F. Discuss the following questions and be prepared to share your answers with your classmates.

1. Which of the three brain structures is largest and heaviest in humans? How do you expect this to compare to the brains of other species?

2. Stroke is a condition caused by clots in blood vessels, leading to a lack of oxygen in nearby tissues and often death or severe damage to those tissues. Strokes in the brain are especially damaging. Based on the information in Figure 13.5 above, a stroke to which part of the brain would be almost certainly deadly?

3. If a stroke damaged part of the cerebellum, what sorts of activities would be affected?

4. Brain damage caused by lack of oxygen to the whole body (for instance, as a result of choking or near drowning) typically affects the cerebrum first, then the cerebellum, and then the brain stem. Why do you think this might be the case?

LAB EXERCISE 13.4

Investigating the General Senses

A. Study reflex reactions.

Most reflex arcs are associated with sensory neurons that transmit pain or body position signals. One commonly tested reflex arc is known as the patellar reflex, which normally is triggered as you begin to lose your balance from an upright position. When you stand, the patellar tendon in your knee is carrying the load of your weight. As you begin to fall, this tendon becomes slack, which triggers the reflex causing your quadriceps (thigh) muscle to contract rapidly, launching your body up in the air. The little hop that results gives your body time to get your feet back under you.

Your patellar reflex may be tested in the course of a physical; any oddities in the reflex might signal neurological trouble.

In this exercise, you will use the rubber hammer available to test the patellar reflex of your lab partner.

1. The subject should sit on the lab bench with legs hanging (not touching the floor). Alternatively, the subject may cross his or her legs so that one is elevated. If you want to measure the reflex response, have another student hold a meter stick perpendicular to your subject's ankle.
2. Locate the patella on the front of your subject's knee, between the bony points of the kneecap and the top of the shinbone.
3. Using the narrower point of the reflex hammer head, tap your partner's patellar tendon firmly.

B. Test sensitivity to touch.

Receptors for touch are not evenly distributed on the surface of the skin. You can observe this with a simple "two-point test." In this test, the subject is touched on various parts of the body by two points simultaneously; the distance between the points at which the subject can distinguish two separate points is a rough measure of the density of touch receptors.

1. Examine the list of body parts in Table 13.1 and rate them according to the density of touch receptors you expect to find.
2. Perform the two-point test on your subject as follows:
    A. Modify a paper clip into a U-shape with equal legs. Wipe the points of the clip with an alcohol-moistened cotton ball to disinfect them. Initially, the distance between the points should be about 1 centimeter.
    B. Your subject should have his or her eyes closed or should be looking away as you perform the test.
    C. For each region of the body listed in Table 13.1, touch the subject's skin lightly with both points of the paper clip. Ask the subjects if they feel one point or two touching their skin. Pick up the points, move them closer together and gently touch the skin of the same area again. The points on the paper clip should be moved closer together until the test subject can no longer distinguish two separate points. You can randomly test the reliability of your subject's responses by touching with only one point occasionally. If they still report two points, you know that they are not a reliable subject for the test.

TABLE 13.1

Body Part	Rating (High, Medium, or Low Number of Receptors)
Forehead	
Cheek	
Lips	
Tongue	
Back of neck	
Back of forearm	
Palm	
Tip of index finger	
Tip of thumb	

D. Measure the distance between the points after the subject can no longer distinguish two points and determine the approximate density of touch receptors by taking the reciprocal of this distance. The reciprocal can be calculated by dividing the measurements into 1. If the two-point discrimination on the cheek is 2.0 mm, the reciprocal would be 0.5. Fill in Table 13.2.

TABLE 13.2

Body Part	Maximum Distance Between Points Where Discrimination Impossible	Reciprocal
Forehead		
Cheek		
Lips		
Tongue		
Back of neck		
Front of neck		
Palm		
Tip of index finger		
Tip of thumb		

3. How accurate were your predictions of density of touch receptors?

4. Compare and contrast two areas that differ in their density of touch receptors. Why do you think there is a difference in sensitivity? Is this a difference individuals are "born with" or do you think insensitivity to touch can be learned?

C. Determine pain perception.

Sensitivity to pain differs markedly among individuals for unknown reasons. Some percentage of these differences are systematic; in other words, certain age groups have higher sensitivity and there are differences in sensitivity between genders.

A simple way to measure pain sensitivity is the "cold pressor test": the number of seconds an individual is able to keep his or her hand completely submerged in an ice-cold water bath is a measure of pain tolerance.

1. Make a hypothesis about differences between genders in pain tolerance. Which gender do you expect to be more tolerant of pain than the other? Be prepared to defend your answer.

2. Make a general prediction about the average results of the cold pressor test for each gender, given your hypothesis.

3. Collect data on cold tolerance among men and women in your class (or recruited from elsewhere) and fill in Table 13.3.

TABLE 13.3   **Number of Seconds Hand Submerged in Ice Water Bath**

Men	Women

Average number of seconds: men _____   women _____

4. Did the results conform to your prediction? Did you support or reject your hypothesis? What further tests might be needed?

5. Were you surprised by the results? Discuss an explanation for them.

D. Study body position.

The general sense that is perhaps most difficult to appreciate is prioperception—the perception of body position. This sense enables us to move effectively in our environment. To demonstrate this "muscle sense," perform the following exercise with a partner.

1. Work in pairs for this exercise.
2. Take two 500 ml beakers, label one A and the other B, and fill both with 250 ml water.
3. Blindfold your subject and place one container in each hand. The weights should feel the same.
4. Take the beakers back. Add 20 ml of water to beaker A, and then ask if the weights feel the same or different. If the subject says "Different," ask which is heavier. If the subject is correct, remove the containers, place them back on the subject's hands, and ask again (to determine whether the subject's answer was simply a lucky guess).
5. If the subject cannot tell the difference between containers or his or her guess was wrong, add another 20 ml of water to beaker A. Continue to test the subject until he or she can detect a difference between container weights.
6. What is the "detection threshold" (the amount at which the subject can distinguish a difference in weight)? Does it differ from one person to the next? Hypothesize about why such differences exist. How could you test your hypothesis?

LAB EXERCISE 13.5

Investigating Sensory Integration

Information from both the general senses and the special senses is integrated by your brain to help you make sense of your surroundings. We are usually unaware of how multiple senses are employed in this integration. This simple demonstration will allow you to see how taste, smell, and vision are integrated when we determine flavor.

A. Work with a partner for this exercise. Obtain a set of envelopes labeled A through H from your lab instructor. Assign four of these envelopes to each lab partner.

B. Put on the blindfold. Plug your nose and have your lab partner feed you a jelly bean from one of your packets. To the best of your ability, identify the flavor of the jellybean and have your partner write your response in Table 13.4. (Please speak softly so that other student pairs are not influenced by your answer.) Repeat this for the other three envelopes.

TABLE 13.4   **Flavor Evaluation**

Envelope Letter	Taste Only	Taste and Sight	Taste, Sight, and Smell

C. For the next round, remove the blindfold, but keep your nose plugged. What do you think the flavor of these candies is now? Write your response on the table.

D. Finally, eat the jellybeans without plugging your nose. Note the flavor on the chart.

E. Discuss the following questions:

1. Did the perceived flavor change over the course of the experiment? Why do you think?

2. Were some flavors easier to diagnose with fewer senses than others?

3. Compare your tables to those of other students in the class. Were some students better than others at discriminating taste with less information? Hypothesize about why there might be a difference. How might you test this hypothesis?

LAB EXERCISE 13.6

Memory and Learning

How memory develops is still a mystery to neuroscientists. What is clear is that a memorable event or fact causes a change in brain biochemistry and that the areas of the brain that are involved in memory and learning can be activated by a variety of sensory triggers. Neuroscientists have also learned a number of different strategies for improving an individual's ability to recall information. You probably have ideas about how best to remember as well. In this exercise, you will design and perform an experiment as a class on the topic of memory and learning.

A. Your lab instructor will read you a list of 20 words. Do not write these down as they are being read! After the list is finished, you should silently write down as many words from this list as you remember, in any order. Do not share your list with your classmates.

B. Your lab instructor will now put the list on the board in the order in which it was read and collect information from you about which words you remembered.

C. Discuss the following questions:

1. Do you see a pattern in the results? Were some words remembered by more students than others? Why do you think there is a difference?

2. Choose one of the hypothesized explanations for why there is a difference in word recall and design an experiment to test your hypothesis.

D. Perform the experiment you have designed. Do the results support or cause you to reject your hypothesis? Explain.

TOPIC 13
_____

POST-LABORATORY QUIZ

THE HUMAN NERVOUS SYSTEM

1. Describe how a nerve impulse is transmitted along a neuron.

2. Describe how a nerve impulse is transferred from one neuron to another.

3. What would happen to nerve transmission if a neurotransmitter remained in a synapse for longer than is typical?

4. Novocaine, a drug used by dentists to "numb" the mouth before pain-causing work, blocks the opening of sodium channels on sensory neurons. How does this reduce pain?

5. Dopamine is a neurotransmitter that stimulates neurons that trigger a sense of well-being and happiness. How would a drug that binds with dopamine receptors in the brain affect an individual in the short term?

6. How does repeated use of a dopamine mimic affect the neurons described previously?

7. List the three major brain regions and briefly describe their function.

8. How is a reflex different from a response that involves brain activity?

9. Describe how several senses interact to produce the sensation of "flavor."

10. Describe how you could test a hypothesis that consuming caffeine interferes with the development of memory.

Ecology and Conservation Biology

Learning Objectives

1. Identify species interactions as mutualism, predation, or competition.
2. Illustrate how changes in the interactions between and among species might directly affect humans.
3. Explain why genetic diversity within an individual can lead to increased fitness of that individual.
4. Define inbreeding depression.
5. Illustrate the process of genetic drift.

Pre-laboratory Reading

As you might guess from the name of the field, the focus of **conservation biology** is on the preservation and restoration of natural environments and habitats. Conservation biologists attempt to bridge the gaps between biology and sociology, environmentalism, law, public affairs, and economics, as well as many other fields of study. This is not an easy task, and often can be controversial. For example, the definition of a "natural habitat" might be different from different perspectives. An environmentalist might feel that "natural" means "free from human influence," whereas an economist might feel that "natural" means "able to produce substantial natural resources for the economy." Thus, these individuals might have completely different visions of what conservation of a particular region should entail. Amidst this controversy, the role of the conservation biologist is mostly to add expertise on the biology and sustainability of **ecosystems**, that is, the organisms in a particular area as well as their nonbiological environment.

A key to understanding ecosystems is an understanding of the science of **ecology**, the study of the interactions between organisms and their environment. In particular, ecologists seek to explain what controls the **distribution** (geographic extent) and **abundance** (total population) of species. A major component of a species' environment is other living species, and interactions among different species can take a number of forms that have differing effects on the ecosystem. When populations of two different species both benefit from each other's presence in a system, the relationship is termed a **mutualism**. The relationship between flowering plants and their pollinators falls into this category, as well as relationships between cleaner fish and their "clients" and fungi and algae within lichens. Two species that use the same or similar sets of resources are said to be in **competition**; in this case, removing one of the two species would cause the remaining species to increase in population size. The interaction between agricultural crops and weeds is a competitive one, as well as interactions between two species of large plant-eating animals on a savannah and those between algae growing on ocean-side rocks. When one species uses another as a food source, the interaction is termed **predation**, with the food source referred to as **prey** and the hunter as **predator**. Changes in the relationships among organisms can have profound effects on an ecosystem. For example, the loss of one species in a mutualism can cause the extinction of the other—this appears to be the case for a species of tree found on the island of Mauritius. The *Calvaria* tree has fruits containing very hard seeds. The seeds will not germinate unless their coats are heavily damaged, and the only way the seed coat becomes damaged is via passage through a large bird's grinding gullet (a digestive organ that functions somewhat like teeth). The only large bird native to Mauritius is the long-extinct Dodo bird. Thus, very few *Calvaria* trees are now found naturally on the island, and those that remain are ancient. (Recent efforts to restore the tree to the island rely on the intervention of humans, who mechanically disrupt the seed coat or feed the fruit to domestic turkeys.)

The decline in a population of a species affects not only interactions within an ecosystem, but also can affect the ability of that species to survive over the long term and to recover from the decline. Species that are **endangered** are typically those that have very small population sizes relative to their historical populations. For example, gray wolves were listed as endangered after their populations in the United States, (excluding Alaska), which once numbered in the tens of thousands, declined to a few hundreds after decades of hunting. Although a few hundred animals is still a lot, these reduced populations face a number of threats. The most difficult of these threats to correct is the loss of genetic diversity caused by **genetic drift**.

Genetic drift refers to *random* changes in **allele frequency** (recall that **alleles** are different forms of the same gene); the frequency of, for instance, allele A1 is the proportion of A1 alleles for gene A in a population. The frequency of an allele can change simply by chance for several reasons, because some deaths occur by accident, or because some individuals with certain alleles happen to not reproduce. Genetic drift can result in large effects on allele frequency in small populations.

One reason that genetic drift is so troublesome is that it leads to a loss of **genetic diversity** in a population. Drift can cause the extinction of particular alleles from a population. Even if an allele currently has little effect on fitness, it might have an effect on the survival and reproduction of individuals under different environmental conditions.

The loss of alleles from a population is also troublesome from an individual's perspective. With fewer alleles present, the chance an individual is **homozygous** for any one gene (that is, possessing two identical alleles for the gene) increases. Homozygosity at a gene locus may be troublesome for two reasons. If an allele codes for a protein that is dysfunctional or functions poorly, an individual homozygous for this allele cannot function properly. Additionally, even if the protein functions well, individuals with two different functional alleles might be successful over a greater range of environments than those who can manufacture just one "version" of the protein. Small populations of a species are also at risk for high homozygosity as a result of **inbreeding**, mating between close relatives. If most of the population is related, the chance that your mate carries similar alleles as you for a number of genes is increased. The decline in fitness that results from an increase in homozygosity in a population is known as **inbreeding depression**.

In this lab, we will investigate the costs of species loss and endangerment to us and to the future of life.

LAB EXERCISE 14.1

Keystone Species

In this exercise, you will take the role of a mammal in the Greater Yellowstone Ecosystem (GYE) to learn how interactions among species can have multiple and sometimes surprising effects on an ecosystem.

A. Your instructor will assign you to a role in the ecosystem game—either gray wolf (gray visors), elk (black visors), or beaver (brown visors).

B. The area designated by the flagging is the GYE. Notice that there are "food items" scattered about the ecosystem: red poker chips indicating aspen, blue chips indicating forage grass.

C. At your instructor's signal, begin foraging for food (and trying to avoid being eaten). Elk feed on both grass and aspen, and can collect both. Beaver feed on aspen only. Wolves feed on elk. Elk and beaver should collect appropriate tokens. Wolves should attempt to tag elk. An elk that is tagged must stop foraging and move off the ecosystem immediately (but should not lose the tokens). Continue foraging until all poker chips are collected or your instructor calls time.

D. Each participant should report the number of food items they consumed, and determine what that means for your next generation:

- Wolves: Consume 1 elk = survive
   Consume 2 elk = survive and have one healthy pup
   Consume 3 or more elk = survive and have two healthy pups
- Elk: Consume 1 aspen or 2 grass = survive
   Consume 2 or more aspen or 4 or more grass = survive and have one healthy offspring
- Beaver: Consume 1 aspen = survive
   Consume 2 or more aspen = survive and have one healthy offspring

E. If you have not consumed enough resources to survive, turn your visor over to your lab instructor. If you have consumed enough resources to reproduce, tell your lab instructor that you need a "baby." The lab instructor will identify one of the nonparticipants (either individuals who did not play in the first round or those who "died" in that round) as the baby and issue the appropriate visor. Elk and beaver should return the tokens to the playing field as your instructor directs

F. Run this simulation four or five more times. Discuss the following questions:

1. Is this ecosystem generally in balance? In other words, are all species of mammal surviving and able to reproduce?

2. One characteristic often seen in predator/prey systems is a cycle where each species experiences regular "boom and bust" populations. Does that seem to be the case with elk and wolves in this ecosystem?

3. How well do you think this simulation approaches reality? What factors are left out of the simulation that might also have an effect on these populations?

G. Now you will investigate the effects of removing a species from the GYE on our simulation. Gray wolves were exterminated from the GYE by the early twentieth century, and were only recently reintroduced (in 1995). To model the effects of this situation, you will play the same "game" minus any wolves. Play five or six rounds.

H. Discuss the following questions and be prepared to share your answers with your instructor and classmates:

1. How did this simulation differ from the previous? Did any species become extinct or become very rare as a result of the loss of wolves? Did any become more common?

2. A keystone is a stone in an archway that helps maintain the shape of the arch. Without the keystone, the arch collapses. Use this definition to describe why wolves are a **keystone species** in the GYE.

3. In the 80 years that gray wolves were missing from the GYE, elk populations increased, and aspen and beaver populations decreased. There were two reasons for this change: changes in the total number of elk and changes in the behavior of elk. Based on how food sources were found in this simulation (which is similar to how they are found in the GYE), what changes in elk behavior are beneficial to aspen?

4. Aspen groves provide homes for a number of species of warblers (small insect-eating song birds), whereas the wetlands produced by beaver activity provide homes for many other plants and animals that depend on these habitats to survive. Write the names of the "players" in this game, as well as the associated species described previously and draw arrows among them to illustrate the feeding connections among these species. (This drawing, as well as the relationships among organisms, is called a **food web**.)

5. Using the figure you drew in question 7, identify species that are in competition with each other, those that are in predator/prey relationships, and those that are mutualists (at least in principle, even if their relationship is not direct).

LAB EXERCISE 14.2

Genetic Drift and Bottlenecks

A **population bottleneck** occurs when a population goes through a rapid and dramatic decrease in size. (This is also called a population crash). In this exercise, you will make and test predictions about the effect of a bottleneck on allele frequency.

A. Work in pairs for this exercise.

B. Obtain a paper bag, a cup of approximately 100 black beads, and a cup of approximately 100 white beads from the center table.

C. Your initial population consists of 100 individuals. The frequency of the "black" allele is 0.5 and the frequency of the "white" allele is 0.5.

D. To model this, place 100 black beads and 100 white beads in the paper bag. (Why do you use 200 total beans to symbolize a population of 100 individuals?)

The protein coded for by the black or white allele has no influence on whether an individual will survive this resource limitation. In other words, right now, the gene is **neutral** with respect to fitness.

E. Imagine that the population in your bag lives in a wetland that is about to be drained for the development of a new housing subdivision. With fewer resources available, only a portion of the population will survive. In other words, the population is about to experience a bottleneck.

F. Predict how the allele frequency in this population will change in response to the bottleneck:

1. Will the frequency of the black and white allele both remain 0.5?

2. Will a bottleneck that reduces the population to 30 individuals have a greater or lesser effect on allele frequencies than a bottleneck that reduces the population to 10 individuals?

G. To test your hypotheses, blindly draw 60 beads from the bag. The beads that are drawn represent the alleles of the 30 survivors of the crash. Count the number of each type of bead and record the data in Table 14.1.

TABLE 14.1

Population Crash to 30 Individuals	Black	White
Number of Beans		
Frequency (Number/60)		

H. Return the 60 beads to the bag and now draw 20 beads (blindly) to symbolize the crash to 10 individuals. Count the number of each type of bead and record the data in Table 14.2.

TABLE 14.2

Population Crash to 10 Individuals	Black	White
Number of Beans		
Frequency (Number/20)		

I. Discuss the following questions and be prepared to share your answers with your instructor and classmates.

1. Was your first prediction correct? Was the frequency after the population crash different than the frequency before the crash?

2. Was your second prediction correct? If not, why not?

3. Did all groups have the same frequency of the black allele after the crash? Why not?

LAB EXERCISE 14.3

Drift in Small Populations

A. Work in pairs for this exercise. You will pool your data with another pair of students.

B. Make hypotheses about the effect of population size and allele frequency on drift.

1. Is an allele more likely to become extinct due to drift in a small population of individuals or a large population of individuals? What is your prediction based on?

2. Is an allele more likely to become extinct if it is initially at a frequency of 0.5 or if it is at a frequency of 0.1?

C. To test these predictions, you and your team will run a series of four simulations of reproduction in a population. Each simulation will follow the same basic pattern, with slightly different numbers inserted at key points. One team should perform simulations I and IV, while the other should perform II and III. The numbers are provided in Table 14.3:

Following is the pattern:

TABLE 14.3

	I. Large Population, Initial Frequency of 0.5	II. Large Population, Initial Frequency of 0.1	III. Small Population, Initial Frequency of 0.5	IV. Small Population, Initial Frequency of 0.1
In (A) spot	30	6	10	2
In (B) spot	30	54	10	18
In (C) spot	60	60	20	20
In (D) spot	30	30	10	10

1. Obtain a paper bag, a cup of 100 black beads, and a cup of 100 white beads. Place (A) black beads and (B) white beads in the paper bag. You will now model the reproduction of this population over five generations.

2. When the adults in this population reproduce, they produce gametes (sperm or eggs) that carry their alleles. If an individual is heterozygote (that is, if it carries one black and one white allele), 1/2 of its gametes will contain the black allele and 1/2 will carry the white allele. We can extend this principle to the entire population as well: If 1/2 of the alleles in a population are black, 1/2 of the gametes produced by the population will carry the black allele. If only 1/10 of the alleles in a population are black, only 1/10 of the gametes produced by that population will carry black alleles. To model reproduction in this population, blindly draw (C) beads from the bag *with replacement*. What this means is that you should draw a bead, note its color, and then return it to the bag before you draw the next one. We replace the beads because although there are only (D) adults (and thus (C) alleles) in the population, each adult can produce hundreds or thousands of gametes, so drawing a gamete containing a particular allele does not reduce the likelihood of drawing a second gamete containing that allele.

3. Record your data in Table 14.4 for the simulation you are currently working on. The frequency here is the frequency of these alleles in generation 2 (the initial population of (C) beads represented generation 1). This frequency in generation 2 might be different than generation 1.

TABLE 14.4

Generation 2	Simulation I	Simulation II	Simulation III	Simulation IV
Number of White Alleles				
Number of Black Alleles				
Frequency of Black Alleles				

4. The numbers in Table 14.4 represent the offspring. This population has a unique allele frequency because it represents an assortment of alleles that were randomly drawn from the parental population. Therefore, to model continued reproduction, you must adjust the population of beads in the paper bag so that it contains the number of black and white beads listed on the table.

5. Now you will repeat the drawing process with this new generation. After you calculate the allele frequency of the offspring, again adjust the numbers of black and white beads in the bag accordingly to model reproduction in the following generation shown in Table 14.5.

TABLE 14.5

Generation 3	Simulation I	Simulation II	Simulation III	Simulation IV
Number of White Alleles				
Number of Black Alleles				
Frequency of Black Alleles				

6. Continue this drawing and adjustment process for 2 more generations (Table 14.6 and Table 14.7).

TABLE 14.6

Generation 4	Simulation I	Simulation II	Simulation III	Simulation IV
Number of White Alleles				
Number of Black Alleles				
Frequency of Black Alleles				

TABLE 14.7

Generation 5	Simulation I	Simulation II	Simulation III	Simulation IV
Number of White Alleles				
Number of Black Alleles				
Frequency of Black Alleles				

7. Repeat steps 1–6 for the other simulation you were assigned.

D. Graph the changes in frequency of the black allele over time for each simulation on the following chart.

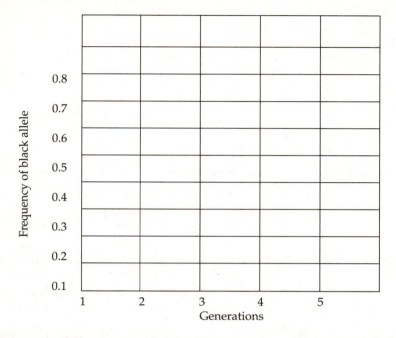

E. Discuss the following questions with your team and be prepared to share your answers with your instructor and classmates:

1. Were your predictions about the effect of population size correct? If not, why not?

2. Was your prediction about the effects of initial allele frequency correct? If not, why not?

3. Did the frequency of the allele change in a steady manner over time? In other words, did it generally increase or generally decrease over the period? Explain this pattern.

4. Did the black or white allele become extinct in any simulations? If not, do you think one would have been lost if we continued this simulation?

5. Consider what we can learn from this simulation about how genetic drift occurs in real populations. How does the initial frequency of an allele affect whether an allele is lost from the population? How does the size of a population affect whether an allele will be lost?

6. Why is it a problem for an endangered species if their population remains small for many generations?

7. After a population has lost alleles due to genetic drift, is there any way to replace these alleles?

LAB EXERCISE 14.4

Biocentrism Quiz

According to many conservation biologists, an important part of encouraging a conservation "ethic" among the nonbiologist public is conservation education. The most effective way that biologists can educate is to simply teach about and make easily accessible information about the natural world.

A. Assess your level of environmental awareness with the following quiz. Work alone. Your lab instructor will provide you with a key and grading guidelines.

1. How many days until the moon is full (plus or minus 2 days)?

2. Name four native edible plants found in this area.

3. From what direction do winter storms come in this area?

4. Where does your garbage go?

5. Name or describe four trees that are native to this area.

6. What primary geological event/process influenced the land form in this area?

7. Name two species that have become extinct in the state in the past 300 years.

8. Were the stars visible last night?

9. What is the primary source of electrical energy for this area?

10. Name four birds that are common in the area.

B. Report your grade on the class chart. After everyone has reported a grade, calculate the average score and note the range.

Average _____

Highest Score _____

Lowest Score_____

C. Discuss the following questions and be prepared to share your answers with your lab instructor and classmates.

1. Did you think the quiz was easy or difficult? Did its level of difficulty surprise you?

2. Consider the average score of the class. How do you think this compares with the average score of an "average citizen" of this region?

3. How important is knowledge of the environment you live in? Are there any negative consequences of not knowing one or more of these basic facts about where you live? What are they?

TOPIC 14
_____

POST-LABORATORY QUIZ

ECOLOGY AND CONSERVATION BIOLOGY

1. Describe the general effect of a keystone species on an ecosystem.

2. Draw a diagram that illustrates the feeding connections (food web) described in the following paragraph:

    Kelp forests grow in the ocean waters immediately off the coasts of western North America. Within these giant stands of brown algae can be found a number of fish species that find shelter and food in the form of the larvae of other sea animals, including other fish, crabs, and sea stars. These smaller animals, in turn, live on smaller algae found in these forests. Sea urchins feed on kelp and can greatly reduce kelp numbers and height. Sea otters feed on sea urchins.

3. Use the preceding diagram to identify competitors, predator/prey pairs, and mutualists.

4. Examine the diagram in question 2. What might happen to the ecosystem if sea otters are removed, as they once where when heavily trapped for their fur?

5. Why do population bottlenecks change allele frequency in a population?

6. Review the following population descriptions. Is one more likely than the other to experience dramatic allele frequency changes as a result of genetic drift? Explain your answer.

    A. Population of 50 individuals living on an isolated island
    B. Population of 250 individuals living on an isolated island

7. Review the following populations. Is one more likely than the other to experience dramatic allele frequency changes as a result of genetic drift? Explain your answer.

   A. Population of 200 individuals, 100 of them breeding
   B. Population of 200 individuals, 10 of them breeding

8. Review the following populations. Is one more likely than the other to experience dramatic allele frequency changes as a result of genetic drift? Explain your answer.

   A. Population of 20 individuals, allele frequency A = 0.9, a = 0.1
   B. Population of 20 individuals, allele frequency A = 0.5, a = 0.5

9. Explain why alleles in low frequency in a population are more likely to become extinct due to genetic drift than alleles in higher frequency, even if the alleles have no current effect on fitness.

10. Describe the costs to individuals of low genetic diversity within a species.

Population Growth and Plant Biology

Learning Objectives

1. Calculate the growth rate of a population from information about births, deaths, and current population numbers.
2. Define carrying capacity and explain how population growth rates change as a population approaches carrying capacity.
3. Investigate the role of water in plant structure and function.
4. Define eutrophication and explain how it occurs.
5. Define biomagnification and explain how energy flow in ecosystems leads to this problem.

Pre-laboratory Reading

The human population of Earth reached 6 billion in 1999 and it continues to increase at a rapid rate. At current rates of growth, the number of people on the planet will exceed 9 billion by the middle of the 21st century. The rapid increase in the human population is a relatively recent phenomenon, as shown in Figure 15.1.

Growth rates in any population are a function of **birth rates** (the number of offspring born per 1,000 members of the population in a given time period) minus **death rates** (the number of deaths per 1,000 members of the population in the same time period). Human populations have exploded in the past 200 years as a result of a dramatic decline in death rates, especially the deaths of infants from infectious diseases. The decline in death rate has led to increased growth rates even as birth rates have remained stable or declined. The result is shown in Figure 15.1—the J-shaped curve is characteristic of **exponential growth**, growth that occurs unchecked, and in proportion to the total population number.

Although exponential growth has been observed in nonhuman populations, none of these populations can grow without limit for very long. All other biological populations eventually run into some environmental limit that sets a ceiling on the number of individuals that can be sustained indefinitely in that

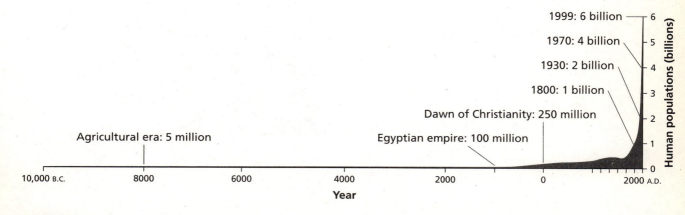

FIGURE 15.1

environment. This ceiling is known as the environment's **carrying capacity** for that population. The carrying capacity might be a function of total food availability, increased rates of disease transmission in larger populations, limits on physical space for individuals, or the like. The growth rate of a population approaching environmental carrying capacity slows down as a result of increasing death rates and/or decreasing birth rates in the population. The change in population size over time in a population limited by the environment can be described by an S-shaped curve, as seen in Figure 15.2.

However, not all populations approach carrying capacity in the gradual manner shown in Figure 15.2. When a population is growing rapidly, it can often overshoot the carrying capacity of the environment. Most often, this type of overshoot is followed by a population "crash," resulting from very high death rates of individuals in an environment where there are too few resources to support them (see Figure 15.3).

Many people who are concerned about the rapid growth of the human population fear that current population levels meet or exceed the carrying capacity of Earth and that humans are headed for a population crash like that shown in Figure 15.3. Critics counter that the human population does not

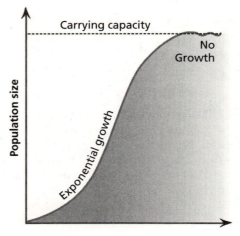

FIGURE 15.2

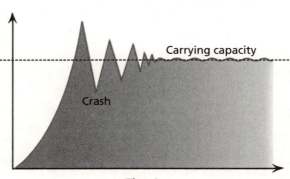

FIGURE 15.3

show any signs of approaching carrying capacity. For instance, death rates continue to decline, food is abundant, and measures of human health are for the most part improving. However, many of these gains in human health and agricultural productivity result from the use of **nonrenewable resources**, primarily **fossil fuels** such as oil, gas, and coal, which will eventually be used up. Whether the eventual result of the loss of these resources will be a crash of the human population is a matter of great debate.

We do know that a major limiting factor for human populations throughout history is food supply. Ultimately, we are all dependent on plant growth for survival. The first major increase in the human population on Earth occurred with the development of **agriculture**, the cultivation and propagation of food plants. The basic premise of agriculture involves strategies that maximize the growth of preferred food plants.

Plants produce food via the process of **photosynthesis**, in which the energy from sunlight is captured in the chemical bonds of **carbohydrates** made from the raw materials of carbon dioxide and water. Maximizing plant growth requires maximizing photosynthesis, so any strategies that increase a plant's exposure to sunlight, carbon dioxide, and water will do this. In addition, plants require **nutrients**—elements that are integral to the pigments and enzymes that perform photosynthesis. Agricultural practices thus also attempt to increase nutrient availability to preferred food plants.

Modern agricultural practices have been extremely successful at maximizing plant growth. Accordingly, the farmers of the world currently produce more food calories than can be consumed by the current human population. These practices include **irrigation** (supplying additional water) and the widespread use of **inorganic fertilizer** (supplying additional nutrients) and **pesticides** (protecting crops from consumption by nonhuman organisms). Unfortunately, each of these practices currently relies heavily on the use of nonrenewable fossil fuels, and each carries with it additional environmental costs. Several of the exercises in this laboratory will explore the environmental costs of modern agriculture.

LAB EXERCISE 15.1

Exploring Population Growth

A. Envision exponential growth.

The dramatic increase in number that occurs when growth is in proportion to population size can be difficult to conceive. This simple exercise will help you envision the power of exponential growth.

1. Obtain a piece of 8.5" x 11" notebook paper. The thickness of this paper (0.1 mm, or 1 ten thousandth of a meter) represents the current population size.

2. Fold the paper in half—this represents a doubling of the population. In one generation, this would be equivalent to a growth rate of 100%. In a population growing at 2% per year (as in the human population), doubling requires about 35 years. The thickness of the folded sheet (2 mm) represents this new population size. Fill in the second row in Table 15.1, indicating the number of layers and the thickness.

TABLE 15.1

Fold Number	Number of Layers	Thickness of Sheaf
0		
1		
2		
3		
4		
5		
6		
7		
8		
9		
10		
11		
12		
13		
14		
15		
16		
17		
18		
19		
20		

3. Fold the paper in half again. The population has doubled again. Fill in the third row in the table.

4. Continue to fold the paper in half. Most people can only fold the paper six or seven times. Continue to fill in the table.

5. Even though you are now at the physical limit of folding, continue to fill in the table until you have calculated the thickness of a piece of paper that has doubled 20 times.

6. Another way to visualize exponential growth in this exercise is to plot a graph of the change in the number of layers over the course of 20 foldings. Obtain a sheet of graph paper from your laboratory instructor. On the horizontal axis of the graph (the X axis), evenly distribute the numbers 1 to 20. Label the vertical (Y) axis of the graph in increments of 1,000 from 0 to 100,000. Plot the number of layers as a function of the fold number on this graph.

B. Calculate population growth rates.

  Examine Table 15.2 and answer the questions. Be prepared to share your answers with your lab instructor and classmates.

TABLE 15.2

Country	Births per 1,000 Population (as of July 2003)	Deaths per 1,000 Population (as of July 2003)	Population Size (as of July 2003)
Australia	12.6	7.3	19,731,984
Czech Republic	9.0	10.7	10,249,216
Norway	12.2	9.7	4,546,123
Panama	20.8	6.2	2,960,784
Tunisia	16.5	5.0	9,924,742
United States	14.1	8.4	290,342,554

1. The growth rate (symbolized by the letter r) of a population is generally determined by subtracting the death rate (d) from the birth rate (b) of that population, divided by the population size. In this case, because birth and death rates are calculated per 1,000 population, we divide by 1,000. Symbolically, this can be written as:

$$r = b - d/1,000$$

Calculate the expected growth rate of the populations in each of the countries in Table 15.3.

TABLE 15.3

Country	Growth Rate
Australia	
Czech Republic	
Norway	
Panama	
Tunisia	
United States	

2. Given the growth rates calculated in question 1, calculate the number of individuals added to (or lost from) each population in 2004 (use Table 15.4). To make this calculation, multiply the 2003 population by the growth rate. Symbolically:

TABLE 15.4

Country	Population Change
Australia	
Czech Republic	
Norway	
Panama	
Tunisia	
United States	

change in population size $= rN$

3. Given this rate of change, what was the population size of each country in 2004 (use Table 15.5)? Symbolically, this calculation can be written as:

$$N_1 = N_0 + \text{change in population size (calculated in question 2)}$$

Where N_1 equals the 2004 population and N_0 equals the 2003 population.

TABLE 15.5

Country	New Population
Australia	
Czech Republic	
Norway	
Panama	
Tunisia	
United States	

4. Use the formula in question 3 to estimate population size in these countries over the next 10 years. Fill in Table 15.6.

TABLE 15.6

Country	2004	2005	2006	2007	2008	2009	2010	2011	2012	2013
Australia										
Czech Republic										
Norway										
Panama										
Tunisia										
U.S.										

5. Review Table 15.6. Examine how the relationship among countries will change over the course of the 10 years. Note that developing countries have higher rates of growth. How is this important? Does this information give you any ideas about strategies to reduce human population growth?

6. The growth rate of a given country actually includes more than just births and deaths—it also includes immigration to the country and emigration from the country. Most governments track this as an aggregate number called the migration rate—immigration minus emigration.

The following formula summarizes the calculation:

$$r = b - d + m/1,000$$

How does the addition of this factor change r in these populations (use Table 15.7?)

TABLE 15.7

Country	Migration Rate	Growth Rate
Australia	4.05	
Czech Republic	0.97	
Norway	2.09	
Panama	0.97	
Tunisia	−0.6	
United States	3.52	

C. Determine the effect of environmental carrying capacity on growth rate.

As a population approaches an environmental limit (that is, the carrying capacity of the environment), its growth rate drops. Ecologists have devised a mathematical formula that summarizes the effect of carrying capacity on growth rate, as follows:

$$\text{Change in population size} = r(K - N/K) * N$$

where r is the intrinsic rate of growth (that is, the rate of growth of the population when resources are essentially unlimited), N is the population size, and K is the maximum population that can be sustained by the environment. The growth rate at any given population size is estimated as:

$$r_{actual} = r_{intrinsic} (K - N/K)$$

Use these equations for the following activities:

1. Given an $r_{intrinsic}$ of 1 and a K of 1,000, calculate the r_{actual} at the population sizes listed in Table 15.8.

TABLE 15.8

N	r_{actual}
10	
50	
100	
200	
500	
700	
900	
1,000	

2. Given an $r_{intrinsic}$ of 1 and a K of 1,000, calculate the population size in the generation immediately following generation N_0 in each row of Table 15.9. Recall that $N_1 = N_0 +$ change in population size.

TABLE 15.9

N_0	N_1
10	
50	
100	
200	
500	
700	
900	
1,000	

3. Describe why population growth rate declines as the population size approaches K in natural populations.

LAB EXERCISE 15.2

Photosynthesis

A. Obtain a leaf from a Variegated Coleus plant. The plant should have been under bright illumination for the past several hours. Draw the leaf here, indicating the pattern of pigmentation on the leaf.

B. Place the leaf in a hot water bath and allow the leaf to simmer for approximately 1 minute. If the leaf contains any purple pigment, keep the leaf in the water bath until all the pigment is gone from the leaf.

C. Using the tongs, transfer the leaf to a beaker containing methanol. Please be sure to replace the watch glass on top of the beaker to minimize evaporation of the methanol. Keep the leaf in the methanol until all of the chlorophyll (the green pigment) has diffused out of the leaf.

D. Using the tongs, transfer the leaf to a petri dish. Note that after this processing, the leaf is flimsy. Carefully flatten the leaf, top side up, in the petri dish.

E. Add several drops of iodine to the petri dish, enough to cover the leaf completely with a thin layer of iodine.

F. You should observe a color change in the leaf. Draw the leaf here, indicating the pattern of pigmentation in the leaf.

G. Answer the following questions and be prepared to share your answers with your laboratory instructor and classmates:

1. Iodine combines with starch to produce a blue-black pigment. Given this information, describe where starch is produced in the leaf. How

does the pattern of starch production correlate to the original pigmentation of the leaf?

2. Starch is the end product of photosynthesis. What does the pattern of starch production and its correlation to leaf pigmentation tell you about the requirements for photosynthesis?

3. Magnesium is a chemical element required for the production of the green pigment, called chlorophyll, in plants. What do you expect would happen to a plant growing in soil that was lacking in magnesium?

LAB EXERCISE 15.3

Water and Plants

A. Discover water transport in plants.

Water is required by plants not just as a raw material for photosynthesis, but also to give the plant shape Adequate water inside plant cells helps provide support for soft tissues, such as leaves. Plants wilt when adequate water is not available, and a wilted plant has a much-reduced rate of photosynthesis relative to a well-watered plant.

Transpiration is the process by which plants transport water from roots to stems and leaves. In this process, water is drawn up the stem as a result of evaporation—much in the same way water is drawn up a straw by suction.

Your lab instructor has placed a plant in dyed water to help demonstrate the process of transpiration. Examine the plant and answer the following questions:

1. Where is the dye located in the plant?

2. The location of the dye tells you something about where on the plant most of the "suction" occurs. Where is this?

3. What do you expect would happen to this plant if you removed it from the beaker of dye?

B. The control of transpiration

Plants do have some ability to control the rate at which water evaporates from the leaves. Pores on the leaves called **stomata (singular: stoma)** allow water to escape (as well as allow carbon dioxide into the leaf for photosynthesis). The size of a stoma is controlled actively by **guard cells**, which surround and define the pore (see Figure 15.4).

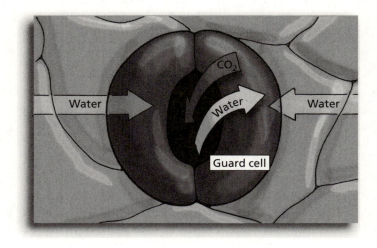

FIGURE 15.4

In this exercise, you will hypothesize about the number of stomata found on leaves of different types, and then examine these leaves for stomata.

1. Examine the leaves available for this investigation. Discuss with your lab partner how they might differ in the concentration of stomata on their surfaces—that is, the number of pores in a given area. Consider the factors that could influence the number of stomata, including potential for water loss and potential for photosynthesis.

2. Develop a hypothesis that you can test about how some or all of the leaves will differ in stomata concentration. Include in your hypothesis your reasoning for WHY you think they will differ in the manner you predict. Write your hypothesis here.

3. To calculate the concentration of stomata on the leaves, follow this protocol:

   • Paint a 1 cm square of the undersurface of the leaf with clear nail polish.

   • Allow the nail polish to dry and then peel it off to produce a "cast" of the leaf. Do not worry if the cast tears; you only need a small amount to make the estimate.

   • Place the cast on a microscope slide, add a drop of water, and cover with a cover slip.

   • Place the slide on the microscope stage and find the cast under lowest power. Center the cast in the field of view and change the magnification on the scope to the highest power (*not* the oil immersion lens, if you have one).

   • Count all of the stomata in the field of view, and then move the slide on the stage slightly so that you are observing a different region of the cast. Count all of the stomata in this field. Continue counting stomata in two more fields of view. Put your data in Table 15.10.

   • Calculate the average number of stomata per field of view

TABLE 15.10

Leaf Type:	Leaf Type:	Leaf Type:
Average number of stomata/ field of view	Average number of stomata/ field of view	Average number of stomata/ field of view

4. Your lab instructor might ask you to present this data in a graphical format.

5. Did the results of your hypothesis test match your predictions? If not, speculate as to why your hypothesis was incorrect. Be prepared to present your hypothesis test to the class.

C. Determine the factors that influence the rate of transpiration.

The rate of transpiration from plants in a natural environment is influenced by a number of factors. In this exercise, you will make and test a hypothesis about which factors increase the rate of transpiration and which decrease the rate of transpiration.

1. Discuss the weather conditions that you think would increase or decrease the rate of transpiration from a plant. Come up with a testable hypothesis that describes a condition that you believe will increase transpiration and one that you believe will decrease transpiration. Write your hypotheses here.

2. Obtain a sunflower seedling, a 1ml glass pipette, a ring stand with ring, and a small section of rubber tubing. You will use these materials to make a "transpiration meter."

3. Submerge the pipette in the water bath and use a pipettor bulb to draw a continuous column of water into the pipette. Do not draw water past the last graduation mark on the pipette. Keep the pipette under water as you complete the rest of the setup.

4. Submerge the cut end of the sunflower seedling into the water bath and use the razor to cut the lower 2 cm of the seedling. Use the rubber tubing as a connector between the narrow end of the pipette to the cut end of the seedling.

5. Remove the entire apparatus from the water bath. Wipe water from the outside of the pipette. If water is dripping from the end of the pipette, the column of water is not continuous—you will need to start over.

6. Wait five minutes to allow the seedling to acclimate to the change. You will now measure the amount of water (in ml) lost from the column in the environmental conditions you specified in your hypothesis.

7. Place your plant in one of the environmental conditions you specified. Record the amount of water lost from the pipette every three minutes for nine minutes total. Calculate the average amount of water lost for a three-minute period and record in Table 15.11.

TABLE 15.11

End of Time Period	ml of Water Lost
3 minutes	
6 minutes	
9 minutes	
Average for 3-minute period	

8. Place your plant in the second of the environmental conditions you specified. After a five-minute acclimation period, begin another nine-minute recording period (use Table 15.12).

TABLE 15.12

End of Time Period	ml of Water Lost
3 minutes	
6 minutes	
9 minutes	
Average for 3-minute period	

9. Answer the following questions and be prepared to share your answers with your lab instructor and fellow classmates:

1. Which condition had the greatest effect on transpiration rate?

2. Does this result support your hypothesis? If not, explain why you think the results were different from what you expected.

LAB EXERCISE 15.4

Eutrophication

Plant growth generally increases in response to the addition of fertilizer; in fact, this is why farmers add fertilizer to their crops and homeowners apply fertilizer to their lawns. It may come as a surprise, therefore, that fertilizer added to fresh water tends to "kill" these bodies of water. The process that results from fertilizer addition to water is known as **eutrophication**.

Eutrophication proceeds as follows: Fertilizer added to water increases the growth of aquatic plants, especially algae. Increased algae growth fuels an increase in the population of decomposers—particularly bacteria in the water. As bacterial activity increases, they begin using up the oxygen that is dissolved in the water to support their own metabolism. Oxygen levels in the water thus decline, and other aquatic organisms perish as oxygen levels reach very low levels. Waterways that have become eutrophic are generally coated with a greenish scum of algae and can experience die-offs of fish and other aquatic creatures.

Eutrophication is especially a problem in lakes that are in highly urbanized or agricultural surroundings. Communities in Minnesota that are concerned about the health of their lakes might have the option of monitoring the rate of eutrophication using remote sensing units deployed in their lakes. The Web site http://wow.nrri.umn.edu/wow contains data collected from a number of Minnesota lakes on oxygen levels and chlorophyll concentration (either measured as total chlorophyll or as turbidity, which is a measure of the cloudiness of the water) gathered by a remote sensor. This exercise uses these data to test one or more hypotheses about the factors influencing eutrophication.

A. Examine the descriptions of the lakes where data are being collected. Pay special attention to the descriptions of the surrounding land uses. Given these uses, which lakes do you think are at highest risk of eutrophication? Which are at lowest risk?

B. Choose two lakes that you suspect have different risks of eutrophication to compare. Obtain the data for a single summer season for the lakes you will be comparing from the WOW Web site. The data relating to eutrophication includes chlorophyll concentration (listed as Chlor µg/L—not available for all lakes), turbidity (listed as Turb NTU), and dissolved oxygen (listed as dO, % sat). Graph the data (for example, dO at 7 meters from June to September) to compare the lakes over the course of a season.

C. Answer the following questions and be prepared to share your answers with your lab instructor and classmates:

1. Do the data support your initial hypothesis? If not, what pattern did you see?

2. How are chlorophyll concentration, turbidity, and dissolved oxygen related to each other?

3. Given the results and the lake descriptions, can you make any suggestions to communities that are struggling with eutrophication in their lakes?

LAB EXERCISE 15.5

Bioaccumulation and Biomagnification

Several of the inputs in modern agricultural practice are materials that are not found in nature. These inputs include insecticides, herbicides, and fungicides.

Many of these chemicals are **persistent**, which means that they do not break down for long periods. Persistent chemicals can **bioaccumulate** (accumulate in biological tissues) and **biomagnify** (become more concentrated in certain organisms) within a polluted environment.

This simple exercise will help demonstrate the process of bioaccumulation and biomagnification.

A. Your instructor will assign you to a particular category of organism. In this simulation, organisms are classified according to their **trophic level**—that is, according to what they eat. Producers are plants or plant-like organisms that perform photosynthesis. Primary consumers consume producers directly; these are also commonly known as herbivores. Secondary consumers consume primary consumers; these are commonly known as carnivores or predators.

B. Each producer will now begin photosynthesizing and producing sugar for its own use, as well as excess sugar that is available for primary consumers. These excess "packets" of sugar are symbolized by chocolate candies. As they are photosynthesizing, they are also bioaccumulating a persistent pollutant (symbolized by the candy wrapper) in their tissues.

C. Producers will use some of this sugar to support their own metabolism. However, even as they use the energy they have produced, they continue to store the persistent pollutant in their tissues.

D. Along come the primary consumers. They will be eating the producers to obtain their "sugar packets" and will pick up the accumulated persistent pollutant as well (that is, all candy wrappers, including the empty ones).

E. Each primary consumer will use half of the remaining energy to support its own metabolism. Again, the persistent chemicals it has ingested will remain in its tissues.

F. Finally, the secondary consumer(s) will eat. Each secondary consumer should obtain the "sugar packets" from primary consumers. While doing this, they will pick up any accumulated persistent pollutants. Each secondary consumer will use half of the energy gained to support its own metabolism, but will retain all of the persistent pollutant.

G. Calculate the average number of wrappers (both empty and containing candy) per individual in each trophic level and enter the data in Table 15.13.

TABLE 15.13

Producers	Primary Consumers	Secondary Consumers

Draw a bar graph of these data, in which each bar represents the average number of wrappers in a given trophic level.

Average number of wrappers per individual

Producers   Primary Consumers   Secondary Consumers

H. Answer the following questions and be prepared to share your answers with your lab instructor and classmates.

1. Describe in your own words why secondary consumers contain much higher concentrations of persistent pollutants than producers or primary consumers.

2. The pesticide DDT is a persistent chemical that was used widely in the United States from the 1940s until the 1970s. DDT was found in very high concentrations in birds that eat large fish, such as bald eagles and osprey. In fact, high levels of DDT in both of these bird species led to dramatic declines in their populations. However, DDT did not have such a dramatic effect on birds that eat small fish, such as common loons, or birds that eat aquatic plants, such as mallards, in the same environments. Use your understanding of biomagnification to explain why bald eagles and osprey were more affected by DDT than loons and mallards.

3. Persistent pollutants also can accumulate in human tissues. One serious concern related to this fact is that breast milk fed to infants can have extremely high levels of these persistent compounds. Again, given your understanding of biomagnification, explain why infants who are breastfed consume higher concentrations of these chemicals than their mothers.

4. Although many persistent pesticides have been banned in the United States, some are still used as pesticides on crops grown in other countries and imported to the United States. Additionally, some persistent pollutants are produced in the United States as a result of the manufacture of pesticides (including pesticides whose use is banned here) and other products. Many of these pollutants have known negative health effects. Should the U.S. government ban the production, sale, and importation of these persistent pollutants, even if such a ban limits the pesticides available to farmers, or increases the costs of goods to consumers? Why or why not? What considerations influence your answer?

TOPIC 15
_____

POST-LABORATORY QUIZ

POPULATION GROWTH AND PLANT BIOLOGY

1. What has caused the dramatic increase in human population over the past 200 years?

2. What is the growth rate of a population where the number of births per 1,000 population is 10.8, the number of deaths per 1,000 population is 7.2, and the migration rate per 1,000 is –2.4?

3. What is the actual growth rate of a population of 900 individuals in an environment where the carrying capacity is 1,100 and the intrinsic growth rate is 1.4?

4. What causes a decline in the growth rate of populations that are approaching the carrying capacity of the environment?

5. Starch is only produced in parts of a plant that contain chlorophyll. What does this information tell you about the role of chlorophyll in a plant?

6. Describe how water moves up the stem of a plant from the roots to the leaves.

7. Name at least two environmental factors that are likely to cause a high rate of transpiration in plants and explain why they do.

8. Describe why increasing the nutrient level in lakes can lead to fish kills.

9. Describe some environmental conditions where lake eutrophication is more likely to occur.

10. Persistent toxic chemicals are found in highest concentrations in organisms at the highest trophic levels. Explain why this is the case.